SCIENCE PLUS®

TECHNOLOGY AND SOCIETY

LEVEL GREEN

TEACHING RESOURCES

Unit 1
Science and Technology

HOLT, RINEHART AND WINSTON
Harcourt Brace & Company

Austin • New York • Orlando • Atlanta • San Francisco • Boston • Dallas • Toronto • London

To the Teacher

This booklet contains a comprehensive collection of teaching resources. You will find all of the blackline masters that you need to plan, implement, and assess this unit. Also included are worksheets that correspond directly to the SourceBook.

Choose from the following blackline masters to meet your needs and the needs of your students:

- **Home Connection** consists of a parent letter designed to get parents involved in the excitement of the *SciencePlus* method. The letter provides parents with a general idea of what you are going to cover in the unit, and it even gives you an opportunity to ask for any common household materials that you may need to accomplish the unit's activities most economically.

- **Discrepant Event Worksheets** provide demonstrations and activities that seem to challenge logic and reason. These worksheets motivate students to question their previous knowledge and to develop reasonable explanations for the discrepant phenomena.

- **Math Practice Worksheets** and **Graphing Practice Worksheets** help fine-tune math and graphing skills.

- **Theme Worksheets** encourage students to make connections among the major science disciplines.

- **Spanish Resources** include Spanish versions of the Home Connection letter, plus worksheets that outline the big ideas and principal vocabulary terms for the unit.

- **Transparency Worksheets** correspond to teaching transparencies to help you reteach, extend, or review major concepts.

- **SourceBook Activity Worksheets** reinforce content introduced in the SourceBook.

- **Resource Worksheets** consist of blackline-master versions of charts, graphs, and activities in the Pupil's Edition.

- **Exploration Worksheets** consist of blackline-master versions of Explorations in the Pupil's Edition. To help students focus on specific tasks, many of these worksheets include a goal, step-by-step instructions, and even cooperative-learning strategies. These worksheets simplify the tasks of assigning homework, allowing opportunities for make-up work, and providing lesson plans for substitute teachers.

- **Unit Activity Worksheet** consists of an activity, such as a crossword puzzle or word search, that provides a fun way for students to review vocabulary and main concepts.

- **Review Worksheets** consist of blackline-master versions of the review materials in the Pupil's Edition, including Challenge Your Thinking, Making Connections, and SourceBook Unit CheckUp.

- **Chapter Assessments** and **End-of-Unit Assessments** provide additional assessment questions. Each assessment worksheet includes two or more Challenge questions that encourage students to synthesize the main concepts of the chapter or unit and to apply them in their own lives.

- **Activity Assessments** are activity-based assessment worksheets that allow you to evaluate students' ability to solve problems using the tools, equipment, and techniques of science.

- **Self-Evaluation of Achievement** gives you an easy method of monitoring student progress by allowing students to evaluate themselves.

- **SourceBook Assessment** is an easy-to-grade test consisting of multiple-choice, true-false, and short-answer questions.

For your convenience, an **Answer Key** is available in the back of this booklet. The key includes reduced versions of all applicable worksheets, with answers included on each page.

Credits: See page 77.

SCIENCEPLUS is a registered trademark of Harcourt Brace & Company licensed to Holt, Rinehart and Winston, Inc.

Printed in the United States of America

ISBN 0-03-095659-5

2 3 4 5 6 7 8 9 021 99 98 97 96

Unit 1: Science and Technology

Contents

▼ *A corresponding transparency is available. See the Teaching Transparencies Cross-Reference on the next page.*

Contents, continued

Worksheet	Page	Suggested Point of Use
Spanish		
Contacto en la casa (Home Connection)	73	Prior to Unit 1
Las grandes ideas (The Big Ideas)	75	Prior to Unit 1/Throughout Unit 1
Vocabulario (Vocabulary)	76	Prior to Unit 1/Throughout Unit 1
Answer Keys		
Chapter 1	77	Throughout Chapter 1
Chapter 2	82	Throughout Chapter 2
Chapter 3	90	Throughout Chapter 3
Unit 1	96	End of Unit 1
SourceBook Unit 1	102	End of SourceBook Unit 1

Teaching Transparencies Cross-Reference

Transparency	Corresponding Worksheet	Suggested Point of Use
Chapter 1		
1: Cracking the Code	Resource Worksheet, p. 7	Lesson 1, Cracking the Code, p. 12
2: Another Cryptogram	Review Worksheet, p. 8	Challenge Your Thinking, p. 16
Chapter 2		
3: Eyeball Benders		Lesson 1, Exploration 3: Eyeball Benders, Tests 1–3, p. 23
4: More Eyeball Benders		Lesson 1, Exploration 3: Eyeball Benders, Tests 1–3, p. 23
5: Be Scientific Follow-Up Chart	Exploration Worksheet, p. 22	Lesson 2, Exploration 4: It's Your Turn to Be Scientific, p. 27
Chapter 3		
6: Graph of Paper Falling Through Air		Lesson 1, Investigating Paper Falling Through Air, p. 41
7: A Hidden Word Puzzle	Unit Activity Worksheet, p. 48	End of Unit 1

Dear Parent,

In the next few weeks, your son or daughter will be introduced to the nature of science and what scientists do. He or she will read about the work of some real scientists and then have the opportunity to act as a scientist by participating in a number of hands-on experiments. By the time the students have finished Unit 1, they should be able to answer the following questions to grasp the "big ideas" of the unit.

1. What does it mean to be scientific? (Ch. 1)

2. What is an observation? an inference? a conclusion? (Ch. 2)

3. How does each of these help people do science? (Ch. 2)

4. How does an investigative question differ from a question that is not investigative? (Ch. 2)

5. How are the words *hypothesis, variable,* and *controlled experiment* related? (Ch. 2)

6. What is the purpose of experiments? (Ch. 2)

7. What are some of the steps in designing experiments? (Ch. 2)

8. How are science and technology related? (Ch. 3)

9. How do science and technology help each other? (Ch. 3)

Listed below are some activities that you may want to do with your son or daughter at home.

• Ask your child what he or she thinks science is all about. Make a mental note of the answer. In a few weeks, ask your son or daughter what he or she now thinks science is all about. Is there a difference between his or her initial and final responses?

• The next time you see an article that is related to science or technology, share it with your child. You may need to explain the article if the material is hard to understand. By sharing this information with your son or daughter, you will be helping him or her see that science is not just for science class. Science is everywhere!

The strength of this science program is that students make discoveries on their own rather than just reading and memorizing the information or having me give the information to them. The students must find the information for themselves by investigating the questions that are asked. There may be times when the students will take projects home. Encourage your son or daughter to find the answers on his or her own.

Sincerely,

The items listed below are materials that we will use in class for the various science explorations of Unit 1. Your contribution of materials would be very much appreciated. I have checked certain items below. If you have these items and are willing to donate them, please send them to the school with your son or daughter by

_____.

- ○ 1-liter bottles and containers
- ○ 5 mL measuring spoons
- ○ adding-machine tape
- ○ aluminum pie plates
- ○ baking powder
- ○ baking soda
- ○ bowls (small)
- ○ buckets
- ○ candles (medium size)
- ○ confectioner's sugar
- ○ cornstarch

- ○ dishwashing detergent
- ○ disposable plates (plastic or plastic-foam)
- ○ eyedroppers
- ○ food coloring
- ○ index cards
- ○ jar lids
- ○ jars (large)
- ○ magazines
- ○ modeling clay

- ○ newspapers
- ○ paper clips
- ○ plaster of Paris
- ○ salt
- ○ shallow plastic dishes, such as margarine containers
- ○ string
- ○ tape (transparent and masking)
- ○ toothpicks
- ○ metal washers (as weights)

Thank you in advance for your help.

Let's Do Some Science! page 9

Your goal	to learn something about science and what scientists do	**Safety Alert!**

You Will Need

- adding-machine tape
- a metric ruler
- a pencil
- transparent tape
- scissors

Making a Möbius Strip

What to Do

Make a Möbius strip by following these directions. Refer back to the illustrations on page 9 of your textbook if necessary.

1. Cut a piece of adding-machine tape about 75 cm long.

2. Twist the piece of adding-machine tape once, and tape the ends together with transparent tape.

Make Some Predictions

1. How many sides does a Möbius strip have? With a pencil, draw a line down the center of the strip. Do not stop where the strip has been taped together.

   _____

   _____

2. How many pieces of paper do you think you will have if you cut along this line with a pair of scissors? Try it.

   _____

   _____

3. How many pieces of paper do you think you will have if you cut once more along the center of the paper strip? Again, try it!

   _____

   _____

4. What does this activity tell you about science and about what scientists do?

   _____

   _____

   _____

Name _____ Date _____ Class _____

Using the Theme of Changes Over Time   Teacher Demonstration

This worksheet is an extension of the theme strategy outlined on page 10 of the Annotated Teacher's Edition. It is designed as an extension to Let's Do Some Science! on page 9 of the Pupil's Edition. The demonstration could also be considered a discrepant-event activity.

Use the activity to show students how scientific understanding evolves over time. This process takes place as scientists accumulate information by investigating, questioning, and evaluating ideas.

Focus question	How do your ideas change as you progress through this exercise?	**Safety Alert!**

You Will Need

- two identical 12 in. helium-quality balloons
- a metal nail file
- an aluminum knitting needle, small diameter (size 3), or a very large upholstery needle
- glycerin
- tissues

Preparation

You may wish to practice this demonstration before performing it in the classroom. Prepare the knitting needle by filing its pointed end to a very sharp point. Next, use a tissue to apply a coat of glycerin to the needle. Prepare a few extra tissues with glycerin on them, and take them with you to class.

Demonstration

1. In front of the class, blow up a balloon and tie the end. Don't inflate the balloon all the way; there should still be a thick portion of rubber at the end after inflating. Repeat with the other balloon.

2. Show students the balloons and the needle, and ask them to predict what would happen if you were to stick the needle into the balloons. *(Expected answer: The balloons will pop.)*

3. Say: Do you mean like this? Then quickly insert the needle through the side of the balloon, as shown below. The balloon should pop loudly.

Chapter 1

4. Then explain to students that you are now going to conduct some magic. With great fanfare, coat the needle with the tissue containing glycerin. Try to do so in a dramatic way that seems to serve no purpose other than being part of your "magic" act.

5. Continue your magic routine by inserting the needle through the bottom (or neck) of the balloon with exaggerated care. Then dramatically push the needle all the way through the balloon and out the opposite (upper) end of the balloon as shown below. The balloon should not pop.

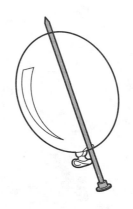

Discussing the Results

Ask students: Can anyone explain what happened? Students will probably comment on the different procedures that were used for each balloon. Ask: Why would the direction in which I poked the balloon with the needle make any difference? Encourage students to examine the balloons up close and to offer an explanation. You may even wish to have students divide into small groups, provide each group with materials, and encourage the students to try to repeat the experiment. Then encourage a class discussion about the successes and failures of each group.

Finally, explain to students that the reason you were able to pierce the second balloon without causing it to pop is related to two facts. First, the air pressure inside the balloon is greater than the air pressure outside the balloon. Second, you pierced the thickest part of the balloon. When you inserted the needle into the neck of the balloon, friction between the rubber of the balloon and the needle formed a seal. For the seal to hold, the rubber had to stretch. On the side of the balloon, where the rubber was not as thick, the rubber could not stretch much further. So the seal could not hold, and the greater air pressure inside the balloon caused it to pop.

Tell students that they will learn more about the science of friction and air pressure in later units. In the meantime, this activity was designed to help them understand that science is self-correcting. If one theory turns out to be wrong, it can be changed. It all begins with observations and predictions. Those predictions can be confirmed or denied through experimentation. If the predictions turn out to be wrong, it's time to develop more theories and conduct more experiments. This process is what science is all about—changes over time!

Name _____ Date _____ Class _____

Making the "Best" Airplane, page 11

An At-Home Activity

1. At home, construct the model airplane shown in the drawing.

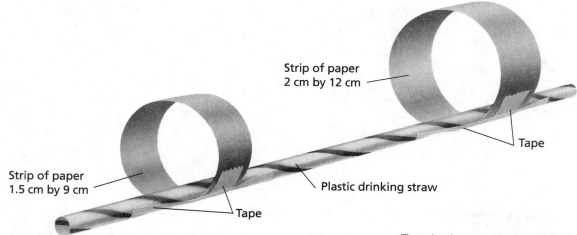

Strip of paper
2 cm by 12 cm

Tape

Strip of paper
1.5 cm by 9 cm

Plastic drinking straw

Tape

Illustration also on page 11 of your textbook

2. Discover the best way to make your plane fly. For example, try placing the loops of paper at different positions along the straw.

3. What else can you try to make the plane fly better?

_____

_____

_____

On a separate sheet of paper, write a report about your findings that you can present to the class. Include a diagram of your most successful design, and include data on the distance your plane traveled.

4. What does this activity tell you about the nature of science?

_____

_____

_____

_____

_____

Cracking the Code, page 12

How is solving a cryptogram similar to what a scientist may do? Try cracking this code, which conceals a definition of science written by the French scientist and mathematician Jules-Henri Poincaré.

To help you, here are two decoded words.

Y V   H X R V M G R U R X
<u>B E</u>   <u>S C I E N T I F I C</u>

H X R V M X V   R H   Y F R O G   F K   D R G S   U Z X G H ,   Z H

_____ __ _____ __ ____ _____ , __

Z   S L F H V   R H   D R G S   H G L M V H .   Y F G   Z

_ _____ __ ____ _____ . ___ _

X L O O V X G R L M   L U   U Z X G H   R H   M L   N L I V

_____ __ _____ __ __ ____

Z   H X R V M X V   G S Z M   Z   S V Z K   L U   H G L M V H

_ _____ ____ _ ____ __ _____

R H   Z   S L F H V .

___ _ _____ .

K L R M X Z I V ,   1 8 8 5

_____ ,   1 8 8 5

After you decode this statement about science, discuss what you think it means with a classmate. Record your ideas here.

_____

_____

_____

_____

Chapter 1
Review Worksheet

Challenge Your Thinking, page 16

1. It's Not a Problem

When was the last time you used scientific thinking (outside of your science classroom)? Describe the circumstances and the type of problem you were trying to solve. Continue in your ScienceLog if necessary.

_____

_____

_____

_____

2. Cryptic Quotes

Solve the cryptogram below to reveal a quote. Here's a clue to help you get started: the word *paths* appears twice. Solving this quote may give you a case of *déjà vu*. Show your solution key in the space below the puzzle.

V NZ NA RKCYBERE. FBZR CNGUF

___ __ _____. _____ _____

GUNG V GENIRY UNIR ORRA

____ _ _____ ____ ____

GENIRYRQ ORSBER. BGURE

_____ _____. _____

CNGUF NER HAPUNEGRQ—

_____ ___ _____—

YRNQVAT GB ARJ VAFVTUGF NAQ

_____ __ ___ _____ ___

GB ARJ QVFPBIREVRF.

__ ___ _____.

Solution key: _____

_____

3. A Clever Solution

Describe the process you used to solve the cryptogram in the previous question. In what ways did your solution method involve scientific thinking?

_____

_____

_____

_____

_____

_____

_____

_____

_____

_____

_____

_____

_____

_____

_____

_____

_____

_____

4. Alert the Media!

Check the daily newspapers for news about science. In your ScienceLog, write a brief report on an article that interests you, and share what you find out with your class.

5. Picture This

In your ScienceLog, create a cover page for this chapter by drawing a diagram or creating a collage of pictures that illustrates what you learned in this chapter.

Name _____ Date _____ Class _____

6. Crazy About Science?

Below is one person's drawing of a scientist.

Illustration also on page 17 of your textbook

What ideas about science and scientists are suggested in the drawing? What ideas presented do you think are true? What ideas do you think are untrue?

_____

_____

_____

_____

_____

_____

_____

_____

**Short
Responses**

1. Which statement do you think most scientists would disagree with?
   Explain your answer in the space provided below.

   a. Science can be fun.   **b.** Scientific ideas never change.

   c. Science can be useful.   **d.** Scientists do not know everything.

   _____

   _____

   _____

   _____

   _____

2. Leonardo is making a cake. Part of the directions read, "Add water and
   mix to desired consistency." The recipe doesn't say how much water
   to use, and Leonardo is unsure about what consistency is desired. He
   wonders how different amounts of water would affect the way the cake
   tastes. Is Leonardo thinking scientifically? Why or why not?

   _____

   _____

   _____

   _____

**Correction/
Completion**

3. Both of the following sentences are false. For each one, write a new, correct
   sentence.

   a. A person can do science only in a laboratory or classroom setting.

   _____

   _____

   _____

   b. Science and curiosity are not related in any way.

   _____

   _____

   _____

Illustration for Interpretation

4. The owner of a fruit market had the names of nine different kinds of fruit posted on a board marked "Today's Specials." A customer accidentally rammed his grocery cart into the sign, and all of the names fell to the floor, as you can see below. Help the owner put the fractured names back together. Then list the fruits in the space provided below.

_____

_____

_____

Describe the process you used to help the owner of the fruit market. Did that process involve scientific thinking? Explain.

_____

_____

_____

_____

Word Usage

5. The term *scientific method* comes from the definition of science as "knowledge" and the Greek word *methodos,* meaning "pursuit" or "going after." Based on this pairing of words, how would you define a scientific method?

_____

_____

_____

CHALLENGE 2

Short Essay

6. How might these people think like scientists?

a. a skateboarder

_____

_____

_____

_____

_____

_____

_____

_____

_____

_____

_____

b. an amusement-ride operator

_____

_____

_____

_____

_____

_____

_____

_____

_____

_____

_____

Chapter 1

EXPLORATION 1

Test Your Powers of Observation, page 20

Cooperative Learning

Group size	3 to 4 students
Group goal	to distinguish between qualitative and quantitative observations
Individual responsibility	Each member of your group should choose a specific role such as materials organizer, safety monitor, reporter, or task leader.
Individual accountability	Each group member should be able to answer question 5 individually.

Safety Alert!

Remember to be careful around open flames!

You Will Need

- matches
- modeling clay
- a candle
- a ruler
- a jar lid or aluminum pie plate
- a watch or clock

What to Do

1. Before lighting the candle, make as many observations about it as possible. Can you list 5? 10? 20? Your time limit is 7 minutes.

   _____

   _____

   _____

   _____

2. Now place the candle on the lid. Modeling clay can be used to hold the candle in place. Light the candle. In the time it takes the candle to burn down (or 10 minutes, whichever comes first), make as many more observations as possible. Use the ruler to help make some observations. Do not, however, burn the ruler or anything other than the candle!

   _____

   _____

   _____

   _____

   _____

Illustration also on page 20 of your textbook

Exploration 1 Worksheet, continued

For Discussion

1. Share your observations with your friends. Did they make observations that you didn't? Classify each observation according to whether it was made by sight, touch, hearing, taste, or smell.

_____

_____

_____

_____

_____

2. Did you make any observations using the ruler? Observations of this type are called **quantitative observations.** Quantitative observations involve measurements and numbers. Perhaps you measured how far the candle burned down in 10 minutes or timed how long it took for the candle to burn down. These are examples of quantitative observations. On the other hand, if you observed the color of the candle, the way the flame flickered, or noted the smell of burning wax, you were making **qualitative observations.** Qualitative observations *do not* involve measurements or numbers.

   Decide which of the observations made by your class were quantitative and which were qualitative. Perhaps you can suggest some more quantitative observations that you could have made.

_____

_____

_____

_____

_____

_____

_____

_____

Chapter 2

Name _____ Date _____ Class _____

3. Reread "A Stranger Has Landed" on page 19 of your textbook, and find statements that express quantitative observations and those that express qualitative observations. List your findings here.

_____

_____

_____

_____

_____

_____

_____

_____

4. Many of your observations of the candle may have described the wax that makes up the candle. Characteristics that help distinguish wax from other materials are called **properties**. For example, wax can be distinguished from ice by its ability to burn: wax will burn; ice will not burn.

a. What other properties of wax would help distinguish a piece of wax from a piece of ice?

_____

_____

_____

_____

b. Actually, ice and wax have many properties in common. For example, they are both solids. Can you think of other properties that they share?

_____

_____

_____

_____

Illustration also on page 20 of
your textbook

Exploration 1 Worksheet, continued

    c. You can make quantitative observations about the properties of wax that would help distinguish it from ice. What observations would you suggest?

_____

_____

_____

_____

    d. Is there a difference between an observation and a property? Explain your reasoning.

_____

_____

_____

_____

5. Write a description of an object as seen through Zed's eyes. Include both quantitative and qualitative observations. Have your classmates guess what you are describing.

_____

_____

_____

_____

_____

_____

_____

_____

_____

_____

Illustration also on page 20 of your textbook

Chapter 2

EXPLORATION 2

Using Observations to Identify Mystery Powders, page 21

Your goal	to use observations to identify three mystery powders	**Safety Alert!**
		Do not try to distinguish substances by taste! Scientists never taste unknown substances because doing so could be dangerous or even deadly.

Three beakers labeled *A, B,* and *C* each contain a white powder. One powder is confectioner's sugar, one is baking powder, and one is plaster of Paris. The following activity will help you determine which is which.

You Will Need

- 3 toothpicks
- 3 jar lids or small bowls
- a small measuring spoon with a capacity of about 5 mL
- samples of mystery powders *A, B,* and *C*
- water
- masking tape
- a pen

What to Do

1. Using the masking tape and the pen, label your lids or bowls *A, B,* and *C.*

2. Put about 5 mL of powder *A* into lid *A,* 5 mL of powder *B* into lid *B,* and 5 mL of powder *C* into lid *C.*

3. Add about 5 mL of water to each lid.

4. Stir each mixture with a toothpick.

5. Observe carefully, and write down your observations. Can you tell from these observations which powder is which?

_____

_____

_____

_____

_____

_____

_____

_____

_____

Exploration 2 Worksheet, continued

If you can tell which powder is which, you are using some information that you already have about the powders and how they should react. If you can't tell, ask your teacher for labeled samples of each substance. Repeat the activity with those substances. Then compare these results to the results you recorded earlier. Now you should be able to identify the mystery powders.

_____

_____

_____

_____

_____

_____

_____

6. When you have finished, wash your hands, and clean up your area.

What is the most important property that enabled you to distinguish the powders from one another?

_____

_____

_____

_____

_____

_____

_____

Chapter 2

Extension

1. You can make this task more challenging by adding more mysterious powders. Two possibilities are baking soda and cornstarch. In addition to testing each powder with water, try using vinegar and an iodine solution. Wear goggles, latex gloves, and an apron for this activity.

2. Record your observations below.

_____

_____

_____

_____

_____

_____

_____

_____

3. Now test a classmate's powers of observation. Prepare a mixture of any two powders. Can he or she name the powders in your mystery mixture?

_____

_____

4. How does this Exploration tell you what science is?

_____

_____

_____

_____

_____

_____

_____

_____

EXPLORATION 3 ···

Eyeball Benders, page 24

Your goal	to test your observational skills	**Safety Alert!**

Test 4

A match is held upright in a test-tube clamp as shown at right and on page 24 of your textbook. Another match is lit and brought near the first one. As the first match bursts into flames, begin your observations.

The rules for Test 4 are as follows:

1. Everyone, except for the person who lights the match, makes observations while seated.

2. Begin making observations at the instant the first match bursts into flame, and end 1 minute after the match goes out.

3. The test may be repeated three times.

Photo also on page 24 of your textbook

Make brief notes about your observations. What did you see? hear? smell? How many observations did you make? How many different observations were made by others? Why didn't everyone make the same observations? What observations did the people sitting near the front make that those farther back missed?

_____

_____

_____

_____

_____

_____

_____

Chapter 2

Name _____ Date _____ Class _____

It's Your Turn to Be Scientific, page 27

Cooperative Learning		Safety Alert!
Group size	3 to 4 students	Be careful around the flame in the Lights Out! activities.
Group goal	to perform an experiment and develop an inference to explain it	
Individual responsibility	Each member of your group should choose a role such as reporter, timer, materials manager, or director.	
Individual accountability	Each group member should be able to explain how the experiment was conducted, what inferences he or she made, what the correct inference was, and what he or she learned from performing the experiment.	

By now you should be an expert at making observations and inferences. In the following activities you will observe some unusual things. For each activity, record two or three observations and at least one inference in the attached Personal Observations and Inferences Chart.

The Dancing Disk

1. Fill an empty glass bottle with cold water.

2. Cut a quarter-sized disk from a plastic or plastic-foam disposable plate.

3. Pour about half of the water out of the bottle.

4. Place the disk on top of the bottle.

5. Grasp the bottle tightly in both hands.

6. Observe what happens to the disk. Make an inference to explain your observations.

The Reappearing Coin

1. Using transparent tape, tape a coin to the bottom of a shallow dish, such as a margarine container.

2. Place the container on the table. Now move backward just far enough so that you cannot see the coin over the rim of the dish.

3. Have a friend pour water into the dish.

4. What do you observe? What inferences can you make from your observations?

Illustrations also on page 27 of your textbook

Exploration 4 Worksheet, continued

The Curious Cup

1. Completely fill a glass with water.

2. Place a piece of paper that is slightly larger than the mouth of the glass over the top.

3. Holding the paper in place, turn the glass upside down over a container or sink, and then remove your hand.

4. What do you observe? What can you infer?

Ice Heist

1. Drop the end of a string into a beaker.

2. Fill the beaker halfway with ice cubes.

3. Sprinkle some salt onto the ice cubes, and wait 30 seconds.

4. Pull on the string. What happens?

5. Make an inference to explain your observations.

Lights Out! #1

1. Stand a candle upright in an aluminum pie plate, holding it in place with modeling clay.

2. Fill the plate halfway with water.

3. Light the candle, and then quickly place an empty jar over the flame.

4. What do you observe? What can you infer?

Lights Out! #2

1. This time, place a candle in a large beaker or jar.

2. Hold the candle in place with a piece of modeling clay.

3. Light the candle.

4. Add 30 mL of baking soda to a 1 L container.

5. Pour in 15 mL of vinegar.

6. Wait 1 minute.

7. Hold the container almost horizontally over the candle, but do not pour the liquid.

8. What happens? What can you infer?

Illustrations also on pages 27 and 28 of your textbook

Chapter 2

Personal Observations and Inferences Chart

Activity	Observations	Inferences
The Dancing Disk		
The Reappearing Coin		
The Curious Cup		
Ice Heist		
Lights Out! #1		
Lights Out! #2		

Exploration 4 Worksheet, continued

Follow-Up

1. For the six activities you just did, combine the observations and inferences that you made in your Personal Observations and Inferences Chart with those of your classmates. Record your answers in the attached Class Observations and Inferences Chart. Place a check mark (✓) beside the best inference.

2. Inferences often raise new questions and suggest more experiments. Consider the Curious Cup activity. What new questions might be raised by it? What experiments might you try in order to answer these questions?

_____

_____

_____

_____

_____

_____

3. Consider this statement: Inferences are not necessarily true. Do you agree or disagree? Use the activities in this Exploration to support your answer.

_____

_____

_____

_____

_____

4. What do you consider to be the value of making inferences?

_____

_____

_____

_____

_____

_____

_____

Chapter 2

Class Observations and Inferences Chart

Activity	Observations	Inferences
The Dancing Disk		
The Reappearing Coin		
The Curious Cup		
Ice Heist		
Lights Out! #1		
Lights Out! #2		

Chapter 2
Review Worksheet

Challenge Your Thinking, page 37

1. Can I Have the Recipe?

Read the two recipes below for making a blueberry pie, and then answer the following questions:

Grandma's recipe	Cookbook recipe
Simply mix three handfuls of berries with two handfuls of sugar, a handful of flour, and a pinch of salt.	Line a 23 cm pie plate with pie crust. Combine in a mixing bowl: 200 mL sugar 60 mL flour 6 mL tapioca 25 mL lemon juice Stir mixture gently into 950 mL fresh blueberries. Pour mixture into pie crust, and dot with 15 mL butter.

a. If you follow Grandma's recipe, will your pie taste the same as hers? Explain.

_____

_____

_____

_____

_____

b. Are Grandma's recipe and the cookbook recipe qualitative or quantitative? Explain.

_____

_____

_____

_____

_____

_____

_____

Chapter 2

2. Who Am I?

On an index card, make a table with the following headings: "Quantitative traits" and "Qualitative traits." Use this table to describe yourself. Choose one person to read the cards (but not the names) to the class so that other students can try to guess who is being described. The best description wins!

3. Thinking Like a Scientist

On page 5 of your textbook, the following statement was made: "This unit is about how you become a scientist each and every day." Do you think this is true? Support your position with examples.

_____

_____

_____

_____

_____

4. Current Events

Choose a newspaper article that has a scientific focus. Underline statements in the article that you think are inferences. Find out if a classmate agrees with you.

5. On Your Own

Try this experiment at home:

- Fill a pie plate halfway with milk.
- Add several drops of different colors of food coloring around the edge of the pie plate.
- Add several drops of dishwashing detergent to the center of the pie plate.
- Observe for 3 minutes.

a. Record as many observations as you can.

_____

_____

_____

_____

_____

_____

_____

Chapter 2 Review Worksheet, continued

b. Suggest several inferences about what you observed.

_____

_____

_____

c. Suggest one investigative question that would help you to discover more about the phenomenon.

_____

_____

_____

d. Suggest a hypothesis based on your question.

_____

_____

_____

6. How Predictable!

In your own words, explain the difference between a prediction and a hypothesis. Use examples if necessary.

_____

_____

_____

_____

_____

_____

_____

_____

_____

_____

_____

Chapter 2

Name _____ Date _____ Class _____

Word Usage   1. Explain how the following pairs of terms are related:

   a. *measurement* and *quantitative observations*

   _____

   _____

   b. *hypothesis* and *cause and effect*

   _____

   _____

   _____

**Illustration for
Interpretation**   2. Examine the situations illustrated below for similarities and differences.

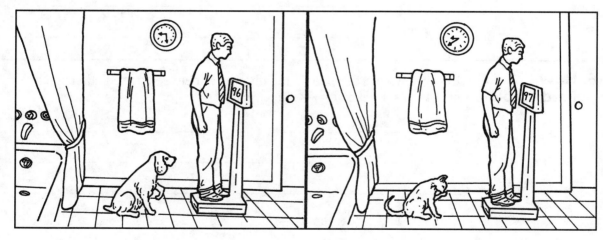

   a. Compare these two situations, and write a quantitative observation
   about what you see.

   _____

   _____

   _____

   b. Compare these two situations, and write a qualitative observation about
   what you see.

   _____

   _____

   _____

c. How might you explain one of the differences you noticed in the drawings?

_____

_____

_____

_____

_____

d. What additional information, if any, would you like to have in order to verify your explanation?

_____

_____

Short Response

3. Read what Jorge and Julie had to say about a problem they noticed, and underline each observation and inference that they made. Then indicate in the right column whether the underlined information is an observation or an inference.

Jorge: Have you noticed how murky and brown the lake is today? _____

Julie: Yes! It looks terrible. Perhaps all the rain we've had recently has washed some extra soil into the lake. _____

Jorge: Perhaps. Or maybe more people have been visiting the lake and leaving their trash behind lately. _____

Julie: I don't think that's the problem because I noticed that the lake was this murky last month after that big rain we had. _____

Jorge: Yes, but that storm happened right after the Fourth of July weekend. This lake gets more visitors that weekend than any other time of year, and I heard that this year more people visited the lake than ever before. _____

Julie: Maybe we're both right. The large number of people using the shorelines of the lake could've trampled a lot of the plants that hold the soil in place. Then the area would've been more vulnerable to erosion. So, when the rains came, more soil was washed into the basin. _____

Jorge: Yes; and maybe the plants that hold the soil in place have still not fully recovered. Until they do recover, I bet we continue to see a murky lake after every rain. _____

Julie: I think we've got ourselves the workings of a pretty good theory here. Let's see if we can go gather some more information from the park ranger! _____

Chapter 2

Chapter 2 Assessment, continued

HRW material copyrighted under notice appearing earlier in this work.

Short Essay

4. Marilyn has three cats: Tiggy, Fluffball, and Boots. She wants to know which brand of cat food they like best. For one week, she feeds Brand A to Tiggy, Brand B to Fluffball, and Brand C to Boots. More of Brand C is eaten that week so Marilyn decides that Brand C must be the best.

Was Marilyn's test fair? Why or why not? If not, what could she do to improve it?

_____

_____

_____

_____

_____

_____

Numerical Problem

5. Kelli has just been to a track meet. She enjoyed watching the hurdles competition and is interested in competing in the event next year, but she is not sure if she can jump high enough. Each hurdle is 91 cm tall. If Kelli is 1.73 m tall, what percentage of her height would she have to jump in order to clear each hurdle? What kind of variables would affect her success? How difficult do you think her attempt would be?

_____

_____

_____

_____

_____

Ball Control   Teacher Demonstration

This tricky problem will help get students ready for the scientific methods they'll be practicing in Chapter 3.

You Will Need

- two plastic cups (about 250 mL or 9 oz.)
- masking tape
- a Ping-Pong ball
- a metric ruler

What to Do

1. Securely tape each cup to a table so that the cups are about 3 cm apart.

2. Put a Ping-Pong ball in one of the cups.

3. Present the following challenge to students: Move the ball from one cup to the other without touching the ball, touching the cups, or using any tools or materials.

4. Ask: Does anyone have any ideas about how this could be done? Record student ideas under the heading Hypotheses on the chalkboard.

5. After you have recorded several hypotheses, ask the class to determine which one seems the most feasible. Then call on volunteers to try that hypothesis.

6. If the experiment isn't successful, ask for feedback from the students. Add new hypotheses or revise existing hypotheses as necessary. You may also wish to call on a student to make notes on the chalkboard about why each hypothesis that fails does so.

7. Call on new volunteers to test a new hypothesis. Continue this process until a solution is reached.

Solution

Blow a short, hard, gust of air at an angle toward the ball. This will cause the ball to move from one cup to the other. Students will probably have to try several angles before being successful. Explain that they are able to move the ball this way because the fast-moving air creates lower air pressure or a partial vacuum on top of the Ping-Pong ball.

Chapter 3

EXPLORATION 1

Air and the "Paper Thing," page 43

Your goal	to investigate how certain types of objects fall through air	**Safety Alert!**

You Will Need

- paper
- paper clips
- scissors
- a metric ruler

What to Do

1. Make a P.T. using the attached diagram.

2. Drop your P.T., and observe its motion.

3. List all of the variables you can think of that could affect its motion.

   _____

   _____

   _____

   _____

4. Write one hypothesis about how you could change the way in which the P.T. falls. Exchange your hypothesis with another group.

   _____

   _____

   _____

5. Design and then try an experiment to gather evidence that either supports or contradicts the hypothesis that you received from the other group.

   _____

   _____

   _____

   _____

   _____

6. Share your plan and the results with the group that gave you the hypothesis.

To make a P.T., cut out the figure below and follow the instructions on each figure. Two figures have been provided so that you can give one to a partner.

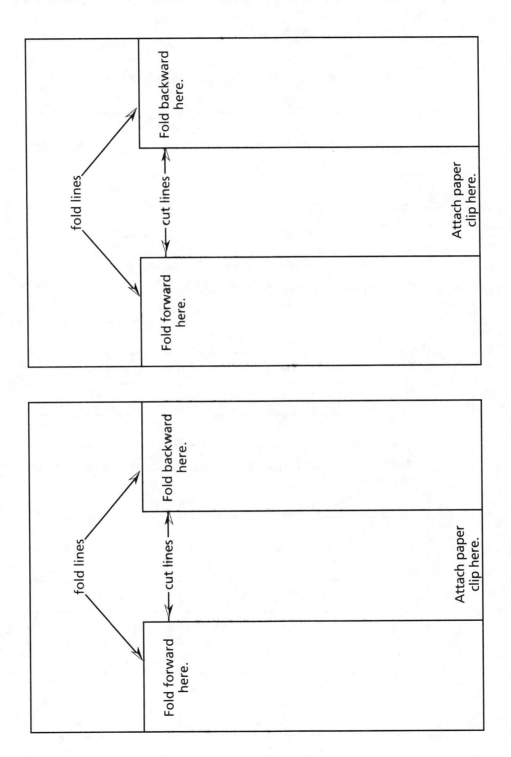

Making a Better P.T.

What makes a better P.T.?

1. In making a better P.T., would you make the same one for each of the following definitions of "better"? Explain.
   - It's one that falls faster.
   - No, slower.
   - It's one that spins more often before hitting the floor.

   _____

   _____

   _____

   _____

   _____

2. What is your definition of a better P.T.?

   _____

   _____

   _____

   _____

3. Make your better P.T., and test it against those of your classmates. Who has the "best" P.T.?

   _____

   _____

   _____

   _____

   _____

   _____

   _____

EXPLORATION 2

Meter Stick Experiments, page 45

Cooperative Learning Activity	
Group size	4 to 5 students
Group goal	to develop a hypothesis to explain reaction times
Individual responsibility	Each member of your group should choose a different role such as recorder, timer, investigator, or orator.
Individual accountability	Each group member should be able to prepare a data sheet that shows the group's results and include a list of any conclusions that were drawn from the results.

Experiment 1   In the first part of this experiment, you are going to compare the reaction times of the people in your group. What factors might have an effect on reaction time?

_____

_____

What to Do

1. Person *A* holds the meter stick between the fingers of person *B*. Person *A* drops the meter stick, and person *B* catches it. The distance that the meter stick falls before person *B* catches it will be a measure of that person's reaction time.

2. In groups of four or five, devise rules that should be followed in order to make this a fair test. These rules should include your controlled variables.

_____

_____

_____

_____

_____

_____

_____

Chapter 3

3. Test the reaction times of the members in your group using your set of rules. Record your answers in the following data chart:

Trial	Distance the meter stick fell

4. Exchange your rules with another group. Try their rules. Do you like their rules better? Answer this question by writing down how you would improve their set of rules.

_____

_____

_____

_____

_____

_____

_____

5. Use your new, revised set of rules to test one or more of the following hypotheses:

• Reaction time is slower when the person is seated.

• Girls have faster reaction times than do boys.

• Right-handed people have faster reaction times than do left-handed people.

On a separate sheet of paper, create a data chart similar to the one above to record your findings.

Exploration 2 Worksheet, continued

Experiment 2

What to Do

1. Hold a meter stick with two fingers, as shown on page 46 of your textbook.

2. Now move your fingers toward the center of the meter stick.

3. Can you move just one finger? _____

4. Try this again, and carefully observe what happens. Where do your two fingers always end up?

_____

_____

_____

_____

A Problem to Solve

1. By adding modeling clay to the meter stick, think of a way for your fingers to end up (a) at the 20 cm mark on the meter stick and (b) at the 40 cm mark on the meter stick.

_____

_____

_____

2. Devise a hypothesis for this experiment.

_____

_____

_____

_____

_____

3. Do your results support your hypothesis? What would be an appropriate conclusion?

_____

_____

_____

_____

_____

Chapter 3

How Long Can It Fly?

Do this activity after completing Lesson 1, which begins on page 40 of your textbook.

Kimiko and Antoine took a rubber-band-powered model plane to a park for the afternoon. They thought it would be fun to experiment with the plane. They took turns winding the rubber band and tossing the plane into the sky. For every one of these tests, they gave the propeller blade a different number of turns. See the illustration below.

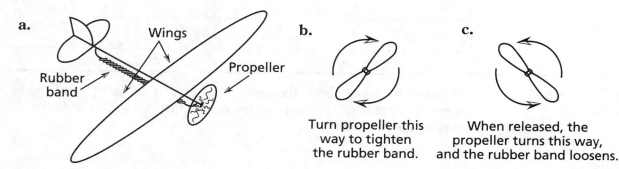

a. Wings, Rubber band, Propeller

b. Turn propeller this way to tighten the rubber band.

c. When released, the propeller turns this way, and the rubber band loosens.

While the plane was airborne, they timed how long it stayed up and recorded that time. They also wrote down the number of turns they gave the propeller. Then they graphed their results below.

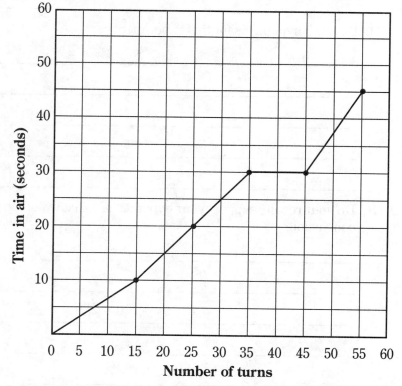

When Antoine tried to turn the propeller more than 55 times, the rubber band broke!

Graphing Practice Worksheet, continued

Think About It

1. The hypothesis that Kimiko and Antoine were testing was that the tighter the rubber band is wound (the more turns it is given), the longer the plane will fly.

2. Antoine and Kimiko were confused because some of their test data did not match their hypothesis, but some of it did. Looking at the graph, can you tell what data caused the problem?

_____

_____

_____

3. Do you think that Kimiko's and Antoine's hypothesis was correct? Are there any other variables not shown on the graph that could have affected their hypothesis?

_____

_____

_____

_____

_____

_____

4. What sorts of variables in the design of the plane affect how long it can stay in the air? Can you think of any changes that might improve the plane's ability to stay airborne?

_____

_____

_____

_____

_____

Extension

Find a partner and try Kimiko's and Antoine's experiment yourself! All you will need is a rubber-band-powered model plane, an open area, some graph paper, and a watch with a second hand. Remember to follow this simple formula: State a hypothesis, design an experiment, collect observations and data, and summarize your findings by drawing a conclusion. Good luck, and happy flying!

Chapter 3

Challenge Your Thinking, page 52

1. Words of Wisdom

Most of our everyday expressions are based on observations. Not all of these observations are accurate. Which of the following expressions do you think are true or partly true? Which could be scientifically tested?

Expression	True or partly true	Testable?
• You can catch a bird by putting salt on its tail.		
• You can't teach an old dog new tricks.		
• You can catch more flies with honey than with vinegar.		
• It's always darkest before the dawn.		
• A watched pot never boils.		
• There is always a calm before a storm.		
• A stitch in time saves nine.		
• A rolling stone gathers no moss.		
• You reap only what you sow.		

Illustration also on page 52 of your textbook

a. Write an investigative statement for four of the expressions above.

_____

_____

_____

_____

_____

_____

_____

_____

_____

_____

Chapter 3 Review Worksheet, continued

b. Design an experiment to test one or more of the expressions.

_____

_____

_____

_____

_____

_____

_____

_____

2. Time Travel

Imagine being transported back in time to 1900, when the first automobiles were traveling the roads. What are some other ways in which the technology of 1900 differs from that of today?

_____

_____

_____

_____

_____

3. Life Imitates Art

Jules Verne, a famous science-fiction writer, wrote about technological advances many years before they were invented. His stories included the submarine, the helicopter, and even the fax machine. Imagine that you are a science-fiction writer and that you are writing about events in the year 2075. What new advances in technology would you include in your story?

_____

_____

_____

_____

_____

Chapter 3

4. Speak for Yourself

The title of Lesson 2 is Technology—Brainchild of Science. In your own words, describe what these words mean to you, and give examples.

_____

_____

_____

_____

_____

_____

_____

_____

_____

5. Draw Me a Map

concept map

Create a concept map to show how the following words are related: *future, knowledge, imagination, science, you, experiments, technology,* and *problems.* To create the map, place these words in circles, arrange them in a logical way, and link them with lines and connecting phrases.

Correction/ Completion

1. The statements below are incorrect or incomplete. Your challenge is to make them correct and complete.

   a. The application of scientific knowledge to solve practical problems and

      make new inventions is _____ .

   b. "I think refrigeration will slow the growth of mold on bread." This is a conclusion about a cause-and-effect relationship.

      _____

      _____

Short Response

Airfoil A   Airfoil B

2. Martha thought that airfoil *A* would have greater lift than airfoil *B*. She is going to test her hypothesis by holding the airfoils in front of an electric fan.

   a. What variable is Martha testing?

      _____

      _____

   b. Name two variables that Martha should control.

      _____

      _____

      _____

Numerical Problem

3. Jeff wants to test three different brands of paper towels to see which absorbs the most water, which is strongest when dry, and which is strongest when wet. He will conduct each test twice to make sure he gets the same results both times. Including the repeated tests, how many tests will Jeff conduct in all?

   _____

   _____

Chapter 3

Illustration for Interpretation

4. Look at the following picture. The car on the left is typical of the cars that were built around 1910. The car on the right is a modern automobile.

Model T   Sports car

a. How are these two kinds of cars similar?

_____

_____

_____

_____

_____

_____

b. How are they different? What kinds of technological advances have been made in cars since 1910?

_____

_____

_____

_____

_____

c. How do you think cars might change in the future?

_____

_____

_____

_____

_____

_____

Chapter 3 Assessment, continued

CHALLENGE 2

Graphing Data

5. The following table compares some common methods of transportation from various times in the nineteenth and twentieth centuries:

Mode of transportation	Average speed (in kilometers per hour)
Stagecoach (1846)	13 km/h
Steam locomotive (1895)	75 km/h
Automobile (present)	100 km/h
Jet airliner (present)	1000 km/h

Using the information in the table, create a bar graph that shows how far each vehicle would travel in 1 hour. Label one axis with the names of the modes of transportation and the other axis with the distance traveled.

a. Using your graph, make one or more observations about transportation changes in the last 150 years.

_____

_____

_____

_____

_____

_____

_____

_____

_____

_____

b. What are some of the benefits of the advances in technology shown on the graph? Do you think there are any disadvantages associated with these changes?

_____

_____

_____

_____

Chapter 3

A Hidden Word Puzzle

Try this activity as you conclude Unit 1.

Unlock the following hidden word, and discover who is known as the "father" of modern science. His name is hidden in the clues below, which describe some of the methods of science that he used. The answers are some of the words that you have learned in Unit 1. The first one is solved for you.

① T E C H N O L O G Y

② _ _ _ _ _ _ _ _ _

③ _ _ _ _

④ _ _ _ _ _ _ _

⑤ _ _ _ _ _ _ _ _

⑥ _ _ _ _ _ _ _

⑦ _ _ _ _ _ _ _ _ _ _ _ _

Clues

In 1609 he used a new example of _____**1**_____, the telescope, to discover that the planet Jupiter has satellites, or moons, revolving around it. Previously, it was assumed that only Earth had a moon.

His discoveries of Jupiter's moons supported a new _____**3**_____ of the solar system with the Sun instead of the Earth at its center. Although this knowledge is taken for granted now, it was a radical idea back then. Previously, having observed that the Sun appeared to move through the sky, people _____**4**_____ that the Sun actually revolved around the Earth.

He did not rely on what others said about the nature of things but made his own careful _____**7**_____ and conducted his own experiments. He was one of the first to do so.

He wondered if all objects, regardless of their mass, fall at the same speed. His _____**6**_____ was that all objects fall at the same rate regardless of their mass.

One story says that he dropped iron balls from the top of the Leaning Tower of Pisa and measured the time it took for them to hit the ground. In this _____**5**_____ experiment, the only _____**2**_____ to change was the mass of the iron balls. His hypothesis was correct.

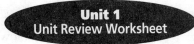

Making Connections, page 54

The Big Ideas   In your ScienceLog, write a summary of this unit, using the following questions as a guide:

1. What does it mean to be scientific? (Ch. 1)

2. What is an observation? an inference? a conclusion? (Ch. 2)

3. How does each of these help people do science? (Ch. 2)

4. How does an investigative question differ from a question that is not investigative? (Ch. 2)

5. How are the terms *hypothesis, variable,* and *controlled experiment* related? (Ch. 2)

6. What is the purpose of experiments? (Ch. 2)

7. What are some of the steps in designing experiments? (Ch. 2)

8. How are science and technology related? (Ch. 3)

9. How do science and technology help each other? (Ch. 3)

Checking Your Understanding

1. Yvonne's group made the statements below as they did the candle activity at the beginning of this unit. Which statements are inferences? Which are observations?

a. The candle is blue.   _____

b. The candle is 5 cm tall.   _____

c. A pool of liquid forms on top of the candle as it burns.   _____

d. This liquid is made of the same substance as the candle.   _____

e. The candle flickers as it burns.   _____

f. Blowing hard on the candle causes it to go out.   _____

g. Blowing hard on a candle causes it to go out because you blow all of the air away from it.   _____

h. Candles need air to burn.   _____

Unit 1

Unit 1 Review Worksheet, continued

2. Below is another cryptogram. Decipher the cryptogram to discover a quotation. Hint: One of the words in the quotation is technology.

V N   V W   N S Z   U H V A V U A   U K   W U Y Z

___   ___   ___   _____   ___   _____

N S M N   O V K Z   U A   Z M F N S   J U E O B

_____   _____   ____   _____   _____

C Z   C Z N N Z F   V K   A U N   K U F

___   _____   ____   ___   _____

W P V Z A P Z   M A B   N Z P S A U U D L .

_____   ____   _____ .

After you crack the code, write a paragraph or two stating whether you agree or disagree with the quotation and why. Use a separate sheet of paper if necessary.

_____

_____

_____

_____

_____

_____

3. Consider the following statement: Scientists discover; inventors invent. Explain in writing what you think this statement means. Indicate whether you agree or disagree and why.

_____

_____

_____

_____

_____

_____

HRW material copyrighted under notice appearing earlier in this work.

4. Once upon a time, there was a young boy who lived in the country. This boy noticed that every morning, just before dawn, the roosters began to crow. He hypothesized that the roosters' crowing caused the sun to rise. Design an experiment to test his hypothesis.

_____

_____

_____

_____

Photo also on page 55 of your textbook

5. If an experiment repeatedly disproves a hypothesis, which of the following actions would be a correct response for a scientist?

a. Ignore the results of the experiment.

b. Keep trying new experiments until the hypothesis is supported.

c. Reject the old hypothesis and form a new one.

d. Conclude that the experiment had some sort of flaw.

Justify your response in writing.

_____

_____

_____

_____

6. (concept map) Complete the concept map below by using the following words: *the world, observing, scientists, inferring, experimenting, variable, hypotheses,* and *models.*

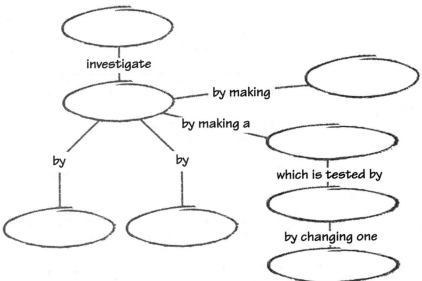

Unit 1

Name _____ Date _____ Class _____

Word Usage

1. The following words belong together. For each pair, discuss why they belong together.

   a. *science* and *discoveries*

   _____

   _____

   b. *observation* and *inference*

   _____

   _____

Correction/ Completion

2. The statements below are incorrect. Your challenge is to make them correct.

   a. In talking about observations, Robert said, "Observations are what you see."

   _____

   _____

   b. In a controlled experiment, every variable is controlled.

   _____

   _____

   c. "Marie has brown eyes and freckles." These are quantitative observations.

   _____

   _____

Short Responses

3. Classify these statements as either observations or inferences.

   a. The red appearance of a sunset is caused by the blanket of air through which the sun is shining. _____

   b. The noonday sun is warmer in the summer than it is in the winter. _____

   c. The sun appears larger as it is setting than it does at noon. _____

   d. The Sun and the Earth are the same age. _____

Unit 1 Assessment, continued

4. Identify the cause and effect in these hypotheses.

   a. The higher the temperature of the water, the faster the eggs will cook.

   _____

   _____

   b. People who drink fluoridated water will have less tooth decay than those who don't.

   _____

   _____

5. You are observing water flowing from a kitchen tap.

   a. Suggest three observations that you could make.

   _____

   _____

   b. Write down one inference.

   _____

   _____

Data for Interpretation

6. Chris and Tammy recorded the following results after flipping a coin 10 times:

Test	Results
1	tails
2	tails
3	heads
4	heads
5	tails
6	tails
7	heads
8	tails
9	heads
10	tails

   a. What percentage of the time did the coin turn up heads? _____

Unit 1

_____

b. If they flipped the coin 100 times, what percentage of the time do you predict the coin would turn up heads?

_____

c. What investigative question could Chris and Tammy have been trying to answer?

_____

_____

_____

CHALLENGE

Short Response

7. Another way of describing technology is to use the term *applied science.* Explain why this is a good synonym for technology.

_____

_____

_____

_____

_____

_____

Data for Interpretation

8. Justine recorded the following data in her ScienceLog.

Amount of sugar in water	Time to boil
0 g	5 min. 53 sec.
6 g	6 min. 5 sec.
12 g	6 min. 17 sec.
18 g	6 min. 29 sec.
24 g	?

a. What investigative question is Justine trying to answer?

_____

_____

_____

b. What variables need to be controlled to make this a fair test?

_____

_____

c. How long do you predict that it would take to boil water with 24 g of sugar in it?

_____

CHALLENGE 2

Short Essay

9. Jill heard an ad on television claiming that a certain shampoo cleaned hair better than any other brand. She decided to do an experiment to test this claim. Describe the experiment she might do. What variables must be controlled to make it a fair test?

_____

_____

_____

_____

_____

_____

_____

_____

_____

_____

_____

_____

_____

_____

_____

Unit 1

**Graph for
Interpretation**

10. The graph below represents data from an experiment. The graph shows how long a music box played after different numbers of turns of the key.

a. What hypothesis do you think is being tested?

_____

_____

_____

_____

_____

b. What should have been done to ensure that the experiment was a fair one?

_____

_____

_____

_____

_____

c. What would be a good conclusion for this experiment?

_____

_____

_____

_____

_____

Thinking Scientifically   Teacher's Notes

Overview	Students make scientific observations and form conclusions based on those observations.

Materials
(per activity station)

- a closed, opaque container with an object inside
- a container half-filled with water
- a mass balance
- a thermometer
- a stopwatch
- a magnifying glass
- a container half-filled with sand
- a magnet
- a measuring cup or graduated cylinder
- a measuring tape
- a flashlight

Preparation

Prior to the assessment, equip student activity stations with the materials needed for each experiment.

Time Required

Allow students 10 minutes to complete Task 1 and 15 minutes to complete Tasks 2 and 3.

Performance

At the end of the assessment, students should turn in the following:

- completed answers for Tasks 1, 2, and 3

Evaluation

The following is a recommended breakdown for evaluation of this Activity Assessment:

- 25% appropriate and logical use of materials and equipment
- 40% quality and clarity of observations
- 35% ability to form conclusion based on observations

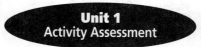

Thinking Scientifically

As a scientist, your job is to answer investigative questions. You get to make observations, form hypotheses, and experiment to test your hypotheses. Show how scientific thinking and sharp science skills can help you accomplish the tasks below.

Before You Begin . . .	As you work through the tasks, keep in mind that your teacher will be observing the following: • how you use the materials and equipment • how clear and complete your observations are • how well you draw conclusions from your observations Now you are ready to tackle the tasks!
Task 1: Mysterious Contents	Without opening the closed container, find out as much as you can about the object inside. You may use any of the materials provided. Record your observations and inferences below.

1. Qualitative observations:

_____

_____

_____

2. Quantitative observations:

_____

_____

_____

3. Inferences:

_____

_____

4. In what ways have you been scientific in your study of the unknown object?

_____

_____

_____

Activity Assessment, continued

Task 2: Sand and Water	Make as many observations as you can about the containers of sand and water.

1. Qualitative observations of container of water:

_____

_____

2. Qualitative observations of container of sand:

_____

_____

3. Quantitative observations of container of water:

_____

_____

4. Quantitative observations of container of sand:

_____

_____

Task 3: Water + Sand = ?	Now pour the water onto the sand and make further observations.

1. Qualitative observations of container of water and sand:

_____

_____

_____

2. Quantitative observations of container of water and sand:

_____

_____

3. Inferences:

_____

_____

_____

4. In what ways have you been scientific in this study?

_____

_____

Unit 1

Unit 1
Self-Evaluation

Self-Evaluation of Achievement

The statements below include some of the things that may be learned when studying this unit. If I have put a check mark beside a statement, that means I can do what it says.

_____ After studying Unit 1, Science and Technology, I can help others to better understand what science is about and what scientists do. (Ch. 1)

_____ I can tell whether a statement is an observation or an inference. (Ch. 2)

_____ I can explain the difference between qualitative and quantitative observations. (Ch. 2)

_____ I have a better idea of how to do a fair test in an experiment than I did before. (Ch. 2)

_____ I can give examples of technology and explain how they are related to science. (Ch. 3)

I have also learned to _____

_____

_____

_____

_____

I would like to learn more about _____

_____

_____

_____

_____

_____

_____

Signature: _____

Name _____ Date _____ Class _____

SourceBook

Flower Power

Complete this activity after reading pages S12–S20 of the SourceBook.

Here's the Problem . . .

Mrs. Greene received a bouquet of white carnations right before she left to go out of town for a 3-day teachers' convention. She asked Craig to take care of the flowers while she was away. "All you have to do," she told him, "is change the water in the vase each day." He agreed enthusiastically (a little *too* enthusiastically, she thought later).

When Mrs. Greene returned, she was surprised to see that the white carnations looked a little different than they did before—they had turned *pink*! When she asked Craig what he had done to the flowers, Craig said that all he did was change the water every day, just like she asked him to do.

Mrs. Greene decided to present the problem to her science class to solve. By using a scientific method that they had learned in school, her class was able to solve the problem. What did they find out? Join with several other students to find an answer. Hint: Craig was telling the truth.

Use the following flowchart and the Experimental Design Form on the next page to help you find the solution.

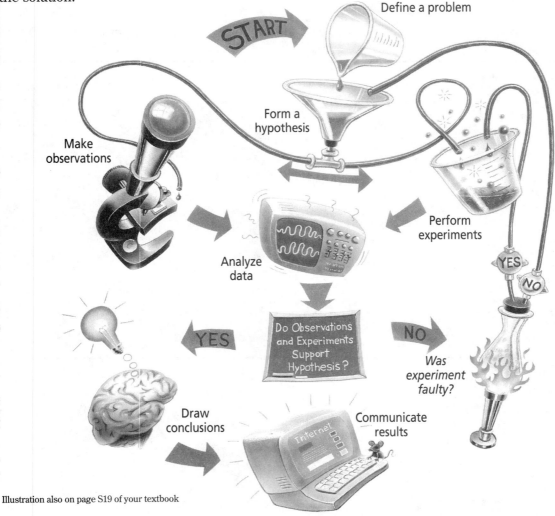

Illustration also on page S19 of your textbook

HRW material copyrighted under notice appearing earlier in this work.

Experimental Design Form

Hypothesis

Plan

Results (Observations)

Conclusion

Name _____ Date _____ Class _____

Unit CheckUp, page S21

Concept Mapping

The concept map shown here illustrates major ideas in this unit. Complete the map by supplying the missing terms. Then extend your map by answering the additional question below.

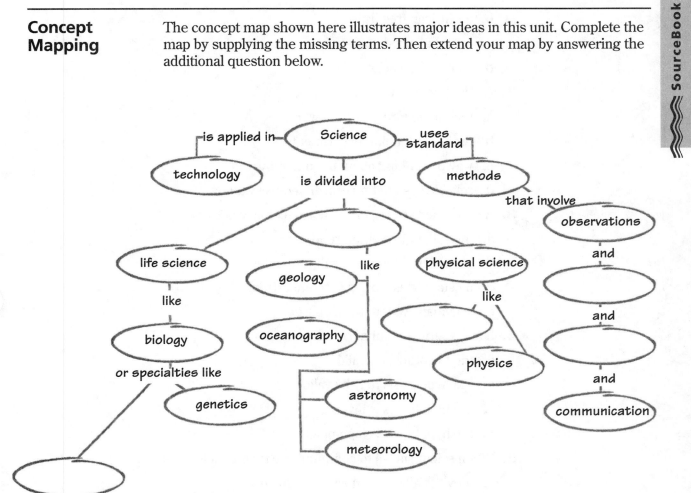

Where and how would you connect the terms *entomology, astrophysics,* and *drawing conclusions?*

Checking Your Understanding

Select the choice that most completely and correctly answers each of the following questions.

1. An Earth scientist would most likely investigate which of the following?

   a. an underwater volcano

   b. an African tree lizard

   c. an unbalanced nuclear force

   d. an insect invasion

2. A scientific method is a

   a. technique for finding answers.

   b. process for doing experiments.

   c. way of thinking about the natural world.

   d. reliable system for recording observations.

3. A *control* for any experiment is necessary because

   a. the scientific method requires one.

   b. it predicts what the results should be.

   c. it establishes a basis for comparison.

   d. scientists want to be in charge.

4. If an experiment supports a hypothesis,

   a. the hypothesis is valid.

   b. the hypothesis may be valid.

   c. the hypothesis is invalid.

   d. the hypothesis is incomplete.

5. When scientists fail to communicate their conclusions,

   a. they get all of the glory for themselves.

   b. their work cannot be considered scientific.

   c. other scientists cannot verify their results.

   d. other scientists must steal their ideas.

Interpreting Graphs

Look at the graph below. What general relationship between temperature and rainfall could you infer from the information given?

_____

_____

_____

_____

At what point in a scientific method would someone produce a graph such as this? Explain.

_____

_____

_____

_____

_____

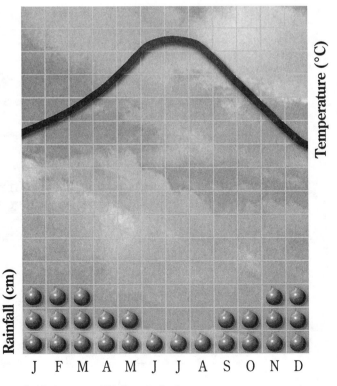

Graph also on page S22 of your textbook

SourceBook Review Worksheet, continued

Critical Thinking

Carefully consider the following questions, and write a response that indicates your understanding of science.

1. Would the question, How can we use hydrogen to run our automobiles? be a question for science or for technology? Explain.

_____

_____

_____

_____

_____

_____

2. Some people argue that astronomy is *not* an Earth science. Why might they say this?

_____

_____

_____

_____

_____

_____

3. Why is it necessary for scientists in different fields, such as biology and physics, to communicate and work together?

_____

_____

_____

_____

_____

_____

SourceBook Review Worksheet, continued

4. Explain why scientists must have at least two samples when they run any experiment.

_____

_____

_____

_____

_____

_____

5. A scientific law might state that the sun will rise tomorrow morning. How might this law be affected if, for some cosmic reason, the Earth stopped rotating?

_____

_____

_____

_____

_____

_____

Portfolio Idea

Imagine that the principal of your school called you to the main office and asked you to explain the meaning of science. What would you say? Think about the science classes you have taken. Would you call what happened in them "science" or not? Make an outline to help you remember the main points you want to bring up. Remember that what you tell the principal may influence how science is taught in your school. Use another piece of paper if necessary.

_____

_____

_____

_____

_____

_____

Name _____ Date _____ Class _____

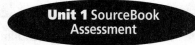

1. Which of the following *best* describes what science is?

   a. Science is the observation of our natural world.

   b. Science is performing experiments to find solutions to problems.

   c. Science is the asking and answering of questions in order to satisfy curiosity about the natural world.

   d. Science is the designing and carrying out of experiments.

2. Which of the following is the *best* description of a scientist?

   a. a person who wears a white coat and works in a laboratory

   b. a doctor

   c. a person who studies endangered animal species in the wild

   d. a person who asks questions and is curious about the natural world

3. If you were a biologist, which of the following would you be *least* likely to study?

   a. life cycles in the tropical rain forest

   b. how plants make food

   c. AIDS research

   d. how earthquakes and volcanic eruptions are predicted

4. A scientific method is only used by scientists.

   a. true   b. false

5. A scientific method is a technique for always getting the right answer to a problem.

   a. true   b. false

6. Which of the following is *not* true of a hypothesis?

   a. Each hypothesis must be tested.

   b. Only one hypothesis can be formed at a time.

   c. A hypothesis should be stated before an experiment is done.

   d. Scientists use observations and experiments to test a hypothesis.

7. The steps of a scientific method are not the same for all scientists.

   a. true   b. false

8. Which of the following has the greatest chance of being incorrect?

   a. a scientific theory   b. a scientific law   c. a hypothesis

9. Scientists observed the following curious events. For each event, give a question that you think scientists might ask about that event.

a. A large community of organisms was observed living around a crack in the ocean floor. This crack was the vent for a hot spring.

_____

_____

b. The eruption of Mount St. Helens on May 18, 1980, turned the surrounding area into a landscape resembling that of the moon.

_____

_____

c. Some rain that falls has an abnormally high acidity. Scientists call this acid rain.

_____

_____

10. What do we call it when scientists apply the knowledge they have gained to improve the quality of life?

a. experimentation   b. technology   c. meteorology   d. science

11. Choose one of the following specific areas of science: botany, genetics, geology, meteorology, or biochemistry. Imagine that you are a scientist in this field. If someone asked you about your work, how would you describe it to him or her?

_____

_____

_____

_____

12. Why do scientists study a specific area of science, like geology, biochemistry, or zoology, instead of studying all areas of science?

_____

_____

_____

_____

_____

SourceBook

13. What are the two main reasons for doing scientific research?

_____

_____

_____

14. Explain why a hypothesis is sometimes called an educated guess.

_____

_____

_____

15. If an experiment is designed to test a hypothesis and it proves that the hypothesis is incorrect, was the experiment a waste of time? Explain your answer.

_____

_____

_____

16. Explain why scientists can change only one variable at a time in an experiment.

_____

_____

_____

17. Based on the results of their study, what did Bekoff and Wells discover about why coyotes run in large packs?

_____

_____

_____

18. Distinguish between a scientific theory and a scientific law.

_____

_____

_____

19. What is meant by the statement, Science moves ahead by correcting errors it made earlier?

_____

_____

_____

20. Form a hypothesis for a question you are curious about, and use the scientific method you learned about in this SourceBook unit to show how you could test that hypothesis.

_____

_____

_____

_____

_____

21. Match the major areas of science on the left with the more specific areas of science on the right. (Each major area will be used twice.)

    a. life science   _____ geology

    b. Earth science   _____ sonar

    c. physical science   _____ meteorology

                            _____ genetics

                            _____ ecology

                            _____ lasers

22. Match the specialized field of science on the left with the correct object of study on the right.

    a. botany   _____ oceans of the world

    b. geology   _____ makeup of matter and its interactions

    c. biochemistry   _____ animals

    d. chemistry   _____ rocky surfaces and the interior of the Earth

    e. oceanography   _____ plants

    f. meteorology   _____ Earth's atmosphere

    g. zoology   _____ energy, its changes, and its relationship to matter

    h. physics   _____ chemistry of life

SourceBook

23. Match the step of a scientific method on the left with the examples on the right.

    a. defining the problem

    b. forming hypotheses

    c. making observations

    d. analyzing data

_____ watching the social behavior of ants

_____ graphing information collected in a recent experiment

_____ a large number of dolphins dying this year

_____ high nitrogen levels in the lake causing fish to die

24. The study of the universe beyond the Earth is known as

_____ .

25. Any information we gather by using our senses is a(n)

_____ .

Estimado padre/madre de familia,

En las próximas semanas, su hijo(a) va a recibir las primeras nociones sobre la naturaleza de la ciencia y lo que hacen los científicos. La clase va a leer cosas relacionadas con el trabajo de verdaderos científicos. Luego, los estudiantes mismos actuarán como científicos, participando directamente en numerosos experimentos. Cuando hayan terminado la Unidad 1, deberán ser capaces de dar respuesta a las siguientes preguntas, para captar las "grandes ideas" de la unidad.

1. ¿Qué significa ser científico? (Cap. 1)

2. ¿Qué es una observación? ¿Una inferencia? ¿Una conclusión? (Cap. 2)

3. ¿En qué forma ayuda cada una de estas cosas a la gente que practica ciencia? (Cap. 2)

4. ¿En qué se distingue una pregunta investigadora de otra que no es investigadora? (Cap. 2)

5. ¿Qué relación hay entre las palabras *hipótesis, variable* y *experimento controlado*? (Cap. 2)

6. ¿Cuál es el propósito de los experimentos? (Cap. 2)

7. ¿Cuáles son algunos de los pasos que se siguen para diseñar experimentos? (Cap. 2)

8. ¿Qué relación hay entre la ciencia y la tecnología? (Cap. 3)

9. ¿Cómo se ayudan mutuamente la ciencia y la tecnología? (Cap. 3)

A continuación se mencionan actividades que, si Ud. quiere, puede practicar con su hijo(a) en la casa.

• Pregúntele a su hijo(a) qué cree que es la ciencia. Tome nota mentalmente de su respuesta. Dentro de unas semanas, pregúntele de nuevo qué cree él o ella que es la ciencia. ¿Hay alguna diferencia entre su respuesta inicial y la final?

• La próxima vez que Ud. vea un artículo relacionado con la ciencia o la tecnología, muéstreselo a su hijo(a). Tal vez necesite explicarle el artículo; si el tema es difícil de entender, tal vez Ud. tenga que explicárselo a su hijo(a). Al compartir con él o ella esta información, Ud. le ayuda a ver que la ciencia no es sólo para la clase de ciencias. ¡La ciencia está en todas partes!

La importancia de esta clase de ciencias está en que los estudiantes hacen sus propios descubrimientos, en vez de simplemente leer y memorizar la información, o de escucharme a mí dándoles la información. Los estudiantes deben encontrar la información por sí mismos, investigando las preguntas que se les hacen. Es posible que haya ocasiones en que su hijo(a) lleve proyectos a la casa. Anime a su hijo(a) a encontrar las respuestas por sí mismo(a).

Atentamente,

Spanish

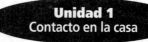

Los materiales que aparecen abajo van a ser usados en clase para varias exploraciones de ciencia de la Unidad 1. Su contribución de materiales va a ser muy apreciada. He marcado algunos de los materiales en la lista. Si usted los tiene y quiere donarlos, por favor mándelos a la escuela con su hijo o hija para el

_____.

○ botellas y recipientes de 1 litro

○ cucharitas para medir (capacidad: aproximadamente 5 mL)

○ papel para máquina de sumar

○ moldes de aluminio para pastel

○ polvo de hornear

○ bicarbonato de sodio

○ tazones (pequeños)

○ cubos

○ velas (tamaño mediano)

○ azúcar de pastelería (impalpable)

○ almidón de maíz

○ detergente para lavar platos

○ platos desechables (de plástico o de espuna de plástico)

○ cuentagotas (goteros)

○ colorante alimentico

○ tarjetas índice

○ tapas para frascos

○ frascos (grandes)

○ revistas

○ plasticina para modelar

○ periódicos

○ clips para papel

○ yeso

○ sal

○ platos de plástico poco profundos, como recipientes para margarina

○ cordón

○ cinta engomada (transparente y tipo "masking")

○ palillos de dientes

○ arandelas de metal (para usar como peso)

Desde ya, le agradecemos su ayuda.

En la Unidad 1, Ciencia y tecnología, vas a investigar la naturaleza de la ciencia y lo que hacen los científicos. Vas a ver que pensar científicamente es un modo de reunir información, organizar lo que se ha reunido, y aprender de tus observaciones. Vas a ver cómo toda la gente usa la ciencia diariamente para encontrarle sentido al mundo que los rodea. Al leer la unidad, trata de responder a las siguientes preguntas. Estas son las "grandes ideas" de la unidad. Cuando puedas contestar estas preguntas, habrás logrado entender bien los principales conceptos de esta unidad.

1. ¿Qué significa ser científico? (Cap. 1)

2. ¿Qué es una observación? ¿Una inferencia? ¿Una conclusión? (Cap. 2)

3. ¿En qué forma ayuda cada una de estas cosas a la gente que practica ciencia? (Cap. 2)

4. ¿En qué se distingue una pregunta investigadora de otra que no es investigadora? (Cap. 2)

5. ¿Qué relación hay entre las palabras *hipótesis, variable* y *experimento controlado*? (Cap. 2)

6. ¿Cuál es el propósito de los experimentos? (Cap. 2)

7. ¿Cuáles son algunos de los pasos que se siguen para diseñar experimentos? (Cap. 2)

8. ¿Qué relación hay entre la ciencia y la tecnología? (Cap. 3)

9. ¿Cómo se ayudan mutuamente la ciencia y la tecnología? (Cap. 3)

Vocabulario

Controlled experiment (34)	**Experimento controlado** un experimento en el cual se controlan todas las variables excepto la que está examinando, para hacer que el experimento sea justo y que los resultados sean confiables
Cryptogram (12)	**Criptograma** un mensaje en código secreto
Hypothesis (35, S14)	**Hipótesis** la explicación de una observación, que puede probarse
Inference (25)	**Inferencia** una conclusión que trata de explicar observaciones que se hacen
Insulin (8)	**Insulina** una sustancia química producida por una pequeña glándula llamada páncreas. Se usa para separar azúcares durante el proceso de digestión, y para tratar la diabetes
Interpret (25)	**Interpretar** explicar el significado de algo
Model (31)	**Modelo** una ilustración, descripción, pequeña reproducción u otra representación que se usa para explicar un objeto, sistema, o concepto
Observation (19, S14)	**Observación** cualquier información que recibamos usando los sentidos
Prediction (33)	**Predicción** adivinar algo antes de que pase
Property (20)	**Propiedad** característica que distingue a una sustancia de otra
Qualitative observation (20)	**Observación cualitativa** observación que no incluye medidas o números
Quantitative observation (20, S17)	**Observación cuantitativa** observación que incluye medidas y números
Science (47, S2)	**Ciencia** conocimiento sobre el mundo natural, derivado de observaciones y experimentos
Supernova (7)	**Supernova** una estrella en explosión, muy grande y muy brillante
Technology (47, S4)	**Tecnología** la aplicación de conocimientos, instrumentos y materiales para resolver problemas prácticos
Variable (34, S16)	**Variable** cualquier factor en un experimento que podría afectar los resultados, y que por eso se examina por separado

Photo/Art Credits

Abbreviated as follows: (t) top; (b) bottom; (l) left; (r) right; (c) center; (bkgd) background.

Photo Credits
Front Cover: (bkgd), Page Overtures. Back Cover: (tl), Tony Stone Images; (bl), Jeff Smith/FotoSmith/Reptile Solutions of Tucson; (bkgd), Page Overtures. Title Page: i (bkgd), Page Overtures; (bl), Jeff Smith/FotoSmith/Reptile Solutions of Tucson; page 21, HRW photo by Sam Dudgeon; 51, H. Reinhard/Okapia/Photo Researchers, Inc.

Art Credits
All work, unless otherwise noted, contributed by Holt, Rinehart and Winston
Page 4, The Mazer Corporation; 5, The Mazer Corporation; 6, John Francis; 10, Valerie Marsella; 14, Reggie Holladay; 16, Reggie Holladay; 17, Stephen Durke/Washington-Artists' Represents, Inc.; 22, Uhl Studio (c), Uhl Studio (b); 23, Uhl Studio (t); Uhl Studio (b-c), Uhl Studio (b); 30, The Mazer Corporation; 32, The Mazer Corporation; 40, The Mazer Corporation; 42, Keith Locke; 45, The Mazer Corporation; 46, The Mazer Corporation; 61, Mark Mille/Sharon Langley Artist Representatives; 65, The Mazer Corporation; 78, John Francis; 80, Valerie Marsella; 82, Reggie Holladay; 83, Stephen Durke/Washington-Artists' Represents, Inc. (t), Reggie Holladay (b); 89, The Mazer Corporation; 90, The Mazer Corporation; 95, The Mazer Corporation; 103, The Mazer Corporation.

Answer Keys

Unit 1: Science and Technology

Contents

Name _____ Date _____ Class _____

Let's Do Some Science! page 9

Your goal	to learn something about science and what scientists do

Safety Alert!
⚠

Making a Möbius Strip

What to Do

Make a Möbius strip by following these directions. Refer back to the illustrations on page 9 of your textbook if necessary.

1. Cut a piece of adding-machine tape about 75 cm long.

2. Twist the piece of adding-machine tape once, and tape the ends together with transparent tape.

You Will Need
• adding-machine tape
• a metric ruler
• a pencil
• transparent tape
• scissors

Make Some Predictions

1. How many sides does a Möbius strip have? With a pencil, draw a line down the center of the strip. Do not stop where the strip has been taped together.

Only one; a pencil line drawn down the center will eventually meet

its starting point.

2. How many pieces of paper do you think you will have if you cut along this line with a pair of scissors? Try it.

Students should end up with one large loop of paper with an

added twist.

3. How many pieces of paper do you think you will have if you cut once more along the center of the paper strip? Again, try it!

By cutting again, students should obtain two interlocking loops

of paper.

4. What does this activity tell you about science and about what scientists do?

Students may suggest that they are doing what scientists do—

making predictions, testing predictions, forming ideas, and

discovering new facts.

Making the "Best" Airplane, page 11

An At-Home Activity

1. At home, construct the model airplane shown in the drawing.

Tape

Strip of paper
2 cm by 12 cm

Plastic drinking straw

Tape

Strip of paper
1.5 cm by 9 cm

Illustration also on page 11 of your textbook

2. Discover the best way to make your plane fly. For example, try placing the loops of paper at different positions along the straw.

3. What else can you try to make the plane fly better?

Students might try varying the width of the loops, the force and angle at which the plane is launched, and the type of straw used.

On a separate sheet of paper, write a report about your findings that you can present to the class. Include a diagram of your most successful design, and include data on the distance your plane traveled.

4. What does this activity tell you about the nature of science?

Sample answer: Science is a process, not just a collection of facts. Scientists experiment and test their ideas to obtain reliable results. Scientists may have to repeat the same experiment many times. Scientists control (keep constant) variables other than the one being tested. Scientists also keep records and communicate their ideas to others.

Cracking the Code, page 12

How is solving a cryptogram similar to what a scientist may do? Try cracking this code, which conceals a definition of science written by the French scientist and mathematician Jules-Henri Poincaré.

To help you, here are two decoded words.

Y V   H X R V M G R U R X
B E   S C I E N T I F I C

H X R V M X V   R H   Y F R O G   F K   D R G S   U Z X G H ,   Z H
S C I E N C E   I S   B U I L T   U P   W I T H   F A C T S ,   A S

Z   S L F H V   R H   D R G S   H G L M V H .   Y F G   Z
A   H O U S E   I S   W I T H   S T O N E S .   B U T   A

X L O O V X G R L M   L U   U Z X G H   R H   M L   N L I V
C O L L E C T I O N   O F   F A C T S   I S   N O   M O R E

Z   H X R V M X V   G S Z M   Z   S V Z K   L U   H G L M V H
A   S C I E N C E   T H A N   A   H E A P   O F   S T O N E S

R H   Z   S L F H V .
I S   A   H O U S E .

K L R M X Z I V ,   1 8 8 5
P O I N C A R É ,   1 8 8 5

After you decode this statement about science, discuss what you think it means with a classmate. Record your ideas here.

Accept all reasonable answers. Students may suggest that the system of finding, organizing, and linking together scientific facts differentiates science itself from just a collection of scientific facts. Similarly, the design and construction of a building distinguish it from a simple collection of stones.

Name _____ Date _____ Class _____

Chapter 1 Review Worksheet, continued

3. A Clever Solution

Describe the process you used to solve the cryptogram in the previous question. In what ways did your solution method involve scientific thinking?

One possible method for solving the puzzle would be to use the clue that

the word *paths* appears twice. The only word with five letters that appears

twice is *CNGUF*; thus, those letters must spell *paths*. By applying these

known letters to replace other letters in the puzzle and making some logi-

cal associations to fill in the remaining letters, the puzzle can be solved.

Some scientific skills involved might be logic, imagination, recalling from

memory, testing, and hypothesizing.

4. Alert the Media!

Check the daily newspapers for news about science. In your ScienceLog, write a brief report on an article that interests you, and share what you find out with your class.

5. Picture This

In your ScienceLog, create a cover page for this chapter by drawing a diagram or creating a collage of pictures that illustrates what you learned in this chapter.

The cover page is intended to allow students to express their own under-

standing of science. Students should include items that show science as a

way of learning about themselves and the world around them.

Name _____ Date _____ Class _____

Chapter 1
Review Worksheet

Challenge Your Thinking, page 16

1. It's Not a Problem

When was the last time you used scientific thinking (outside of your science classroom)? Describe the circumstances and the type of problem you were trying to solve. Continue in your ScienceLog if necessary.

Answers will vary, but encourage students to compare their actions with

the actions of the scientists discussed in the textbook.

2. Cryptic Quotes

Solve the cryptogram below to reveal a quote. Here's a clue to help you get started: the word *paths* appears twice. Solving this quote may give you a case of *déjà vu*. Show your solution key in the space below the puzzle.

V N Z N A RKCYBERE. FBZR CNGUF
I AM AN EXPLORER. SOME PATHS

GUNG V GENIRY UNIR ORRA
THAT I TRAVEL HAVE BEEN

GENIRYRQ ORSBER. BGURE
TRAVELED BEFORE. OTHER

CNGUF NER HAPUNEGRQ—
PATHS ARE UNCHARTED—

YRNQVAT GB ARJ VAFVTUGF NAQ
LEADING TO NEW INSIGHTS AND

GB ARJ QVFPBIREVRF.
TO NEW DISCOVERIES.

Solution key: A = N; B = O; C = P; D = Q; E = R; F = S; G = T;
H = U; I = V; J = W; K = X; L = Y; M = Z

Short Responses

1. Which statement do you think most scientists would disagree with? Explain your answer in the space provided below.

   a. Science can be fun.   b. Scientific ideas never change.

   c. Science can be useful.   d. Scientists do not know everything.

   Most scientists would probably disagree with (b) because much of science is based on theories. Historically, many scientific theories have been disproved or changed in some way.

2. Leonardo is making a cake. Part of the directions read, "Add water and mix to desired consistency." The recipe doesn't say how much water to use, and Leonardo is unsure about what consistency is desired. He wonders how different amounts of water would affect the way the cake tastes. Is Leonardo thinking scientifically? Why or why not?

   Yes, Leonardo is thinking scientifically. He formulated a valid question that could be tested by conducting an experiment.

Correction/ Completion

3. Both of the following sentences are false. For each one, write a new, correct sentence.

   a. A person can do science only in a laboratory or classroom setting.

      Sample answer: A person can do science anywhere; he or she must simply be able to make observations, hypothesize about those observations, and test his or her ideas with experimentation.

   b. Science and curiosity are not related in any way.

      Science and curiosity are related because science helps people find answers to things that they are curious about.

SCIENCEPLUS • LEVEL GREEN   11

---

6. Crazy About Science?   Below is one person's drawing of a scientist.

Illustration also on page 17 of your textbook

What ideas about science and scientists are suggested in the drawing? What ideas presented do you think are true? What ideas do you think are untrue?

Students are expected to dispel misconceptions about science and scientists. This includes the stereotype of scientists as social misfits or mad geniuses. Ideal answers will acknowledge that scientists often do experiments in laboratories, keep records of data, take measurements, and label specimens, but that science involves more than just experiments and is not limited to laboratories.

10   UNIT 1 • SCIENCE AND TECHNOLOGY

Name _____ Date _____ Class _____

Chapter 1 Assessment, continued

Illustration for Interpretation

4. The owner of a fruit market had the names of nine different kinds of fruit posted on a board marked "Today's Specials." A customer accidentally rammed his grocery cart into the sign, and all of the names fell to the floor, as you can see below. Help the owner put the fractured names back together. Then list the fruits in the space provided below.

NG   PLES   PL   UMS   LES

AP   GRA   ONS   CHES

PE   PES   ORA   PEA

WA   TER   MEL   PI

ALOU   PES   ES   CANT   APP

NE

Oranges, apples, pineapples, peaches, pears, grapes, cantaloupes,

watermelons, plums

Describe the process you used to help the owner of the fruit market.
Did that process involve scientific thinking? Explain.

**The following scientific skills may have helped students solve the
problem: logic, imagination, recalling from memory, testing, and
hypothesizing.**

CHALLENGE 1
Word Usage

5. The term *scientific method* comes from the definition of science as "knowledge" and the Greek word *methodos*, meaning "pursuit" or "going after." Based on this pairing of words, how would you define a scientific method?

Sample answer: A scientific method involves the pursuit of knowledge;

it is a method for gaining knowledge about the world around us.

Name _____ Date _____ Class _____

Chapter 1 Assessment, continued

6. How might these people think like scientists?

a. a skateboarder

Sample answer: A skateboarder might observe several locations for a specific type of surface that would provide an appropriate area for skateboarding. For example, he or she may want a smooth surface to develop speed and balance or a more challenging surface, such as one with hills and slopes, to improve style and technique. On the other hand, the skateboarder would probably avoid dirt, gravel, and other rough surfaces that prevent the smooth rolling of the skate-board's wheels. Once the skateboarder observed several places, he or she would probably hypothesize about which would best suit his or her needs and then test the location by skateboarding there. By extending his or her hypothesis, the skateboarder might develop ramps of various heights and lengths.

b. an amusement-ride operator

Sample answer: An amusement-ride operator might observe that a ride does not perform as well under some circumstances as it does under others. For example, the operator may observe that the ride is difficult to start and more likely to stall after a rainstorm. He or she might hypothesize that water gets into the electric components of the ride, keeping the independent parts from operating as they should. He or she might then devise a method for drying out those parts after a rainstorm, test that method, and then propose a solu-tion for improving the ride. The operator might also listen to the riders for screams or laughter. He or she could then operate the ride at various speeds to give the riders the best ride.

CHALLENGE 2
Short Essay

Name _____ Date _____ Class _____

Exploration 1 Worksheet, continued

For Discussion

1. Share your observations with your friends. Did they make observations that you didn't? Classify each observation according to whether it was made by sight, touch, hearing, taste, or smell.

   Observations should be classified according to the sense used.

   _____

   _____

   _____

2. Did you make any observations using the ruler? Observations of this type are called **quantitative observations.** Quantitative observations involve measurements and numbers. Perhaps you measured how far the candle burned down in 10 minutes or timed how long it took for the candle to burn down. These are examples of quantitative observations. On the other hand, if you observed the color of the candle, the way the flame flickered, or noted the smell of burning wax, you were making **qualitative observations.** Qualitative observations *do not* involve measurements or numbers.

   Decide which of the observations made by your class were quantitative and which were qualitative. Perhaps you can suggest some more quantitative observations that you could have made.

   Quantitative observations may include length of the flame, length of the candle, length of the wick, and rate of burning. Qualitative observations may include the color of the candle, the smell of burning wax, and the way the flame flickered.

   _____

   _____

   _____

Name _____ Date _____ Class _____

EXPLORATION 1

Test Your Powers of Observation, page 20

Cooperative Learning

Group size	3 to 4 students
Group goal	to distinguish between qualitative and quantitative observations
Individual responsibility	Each member of your group should choose a specific role such as materials organizer, safety monitor, reporter, or task leader.
Individual accountability	Each group member should be able to answer question 5 individually.

Safety Alert!

Remember to be careful around open flames!

You Will Need

- matches
- modeling clay
- a candle
- a ruler
- a jar lid or aluminum pie plate
- a watch or clock

What to Do

1. Before lighting the candle, make as many observations about it as possible. Can you list 5? 10? 20? Your time limit is 7 minutes.

   Observations will vary but may describe the candle's color, size, and shape.

   _____

   _____

2. Now place the candle on the lid. Modeling clay can be used to hold the candle in place. Light the candle. In the time it takes the candle to burn down (or 10 minutes, whichever comes first), make as many more observations as possible. Use the ruler to help make some observations. Do not, however, burn the ruler or anything other than the candle!

   Observations will vary but may include length of the flame, length of the candle, length of the wick, rate of burning, the color of candle, the smell of burning wax, and the way the flame flickered.

   _____

   _____

Illustration also on page 20 of your textbook

Name _____ Date _____ Class _____

Exploration 1 Worksheet, continued

3. Reread "A Stranger Has Landed" on page 19 of your textbook, and find statements that express quantitative observations and those that express qualitative observations. List your findings here.

Quantitative observations include 198 cm, 3.5 years (or light-years), and 6 fingers. Qualitative observations include red and yellow flowers, green trees, blue sky, and blue skin.

4. Many of your observations of the candle may have described the wax that makes up the candle. Characteristics that help distinguish wax from other materials are called **properties**. For example, wax can be distinguished from ice by its ability to burn: wax will burn; ice will not burn.

a. What other properties of wax would help distinguish a piece of wax from a piece of ice?

Sample answers: Wax can be scratched easily with your fingernail, but ice is harder to scratch; wax does not melt when placed on your desk, but ice does; and wax is not cold to the touch, but ice is.

b. Actually, ice and wax have many properties in common. For example, they are both solids. Can you think of other properties that they share?

Sample answer: Both melt when heated; both float on water; both break easily; and both stay solid below 0°C.

Illustration also on page 20 of your textbook

Name _____ Date _____ Class _____

Exploration 1 Worksheet, continued

c. You can make quantitative observations about the properties of wax that would help distinguish it from ice. What observations would you suggest?

Sample answer: the melting point of each and the mass of a certain volume of each

d. Is there a difference between an observation and a property? Explain your reasoning.

Yes, there is a difference. An observation is something you perceive about an object, but a property is something you use to distinguish it from other objects.

5. Write a description of an object as seen through Zed's eyes. Include both quantitative and qualitative observations. Have your classmates guess what you are describing.

Answers will vary depending on the object chosen, but descriptions should be clear and thorough and should include both quantitative and qualitative observations.

Illustration also on page 20 of your textbook

≋ **Answers • Chapter 2**

Exploration 2 Worksheet, continued

If you can tell which powder is which, you are using some information that you already have about the powders and how they should react. If you can't tell, ask your teacher for labeled samples of each substance. Repeat the activity with those substances. Then compare these results to the results you recorded earlier. Now you should be able to identify the mystery powders.

The substance that dissolves in water is confectioners' sugar; the substance that fizzes in water is baking powder; and the substance that feels warmer after adding water and eventually hardens plaster of Paris.

_____

_____

6. When you have finished, wash your hands, and clean up your area. What is the most important property that enabled you to distinguish the powders from one another?

   The way the powders react with water is the most important property for distinguishing them.

   _____

   _____

   _____

   _____

Chapter 2
Exploration Worksheet

EXPLORATION 2

Using Observations to Identify Mystery Powders, page 21

Your goal	to use observations to identify three mystery powders

Safety Alert!

Do not try to distinguish substances by taste! Scientists never taste unknown substances because doing so could be dangerous or even deadly.

Three beakers labeled *A*, *B*, and *C* each contain a white powder. One powder is confectioner's sugar, one is baking powder, and one is plaster of Paris. The following activity will help you determine which is which.

You Will Need

- 3 toothpicks
- 3 jar lids or small bowls
- a small measuring spoon with a capacity of about 5 mL
- samples of mystery powders *A*, *B*, and *C*
- water
- masking tape
- a pen

What to Do

1. Using the masking tape and the pen, label your lids or bowls *A*, *B*, and *C*.

2. Put about 5 mL of powder *A* into lid *A*, 5 mL of powder *B* into lid *B*, and 5 mL of powder *C* into lid *C*.

3. Add about 5 mL of water to each lid.

4. Stir each mixture with a toothpick.

5. Observe carefully, and write down your observations. Can you tell from these observations which powder is which?

   Students should notice that one of the substances dissolves in water, one fizzes in water, and one feels warmer after adding water and will eventually harden.

   _____

   _____

   _____

   _____

Name _____ Date _____ Class _____

Exploration 2 Worksheet, continued

Extension

1. You can make this task more challenging by adding more mysterious powders. Two possibilities are baking soda and cornstarch. In addition to testing each powder with water, try using vinegar and an iodine solution. Wear goggles, latex gloves, and an apron for this activity.

2. Record your observations below.

Students should observe that vinegar causes both the baking soda and

the baking powder to bubble and that iodine turns both cornstarch and

baking powder black (indicating that both contain starch).

_____

_____

_____

3. Now test a classmate's powers of observation. Prepare a mixture of any two powders. Can he or she name the powders in your mystery mixture?

_____

_____

4. How does this Exploration tell you what science is?

Students might respond that Exploration 2 helps to explain what science

is because it encourages paying attention to procedures, observing

carefully, recording observations, drawing conclusions, repeating experi-

ments, and drawing more conclusions.

_____

_____

_____

_____

_____

Name _____ Date _____ Class _____

Chapter 2
Exploration Worksheet

EXPLORATION 3

Eyeball Benders, page 24

Your goal	to test your observational skills

Safety Alert!

Test 4

A match is held upright in a test-tube clamp as shown at right and on page 24 of your textbook. Another match is lit and brought near the first one. As the first match bursts into flames, begin your observations.

The rules for Test 4 are as follows:

1. Everyone, except for the person who lights the match, makes observations while seated.

2. Begin making observations at the instant the first match bursts into flame, and end 1 minute after the match goes out.

3. The test may be repeated three times.

Make brief notes about your observations. What did you see? hear? smell? How many observations did you make? How many different observations were made by others? Why didn't everyone make the same observations? What observations did the people sitting near the front make that those farther back missed?

Answers should describe the burning of the matches in as much detail as

possible, using sight, hearing, and smell. Students closer to the demonstra-

tion should notice more details than those farther away.

_____

_____

_____

_____

_____

Photo also on page 24 of your textbook

≋ Chapter 2

Name _____ Date _____ Class _____

Exploration 4 Worksheet, continued

Follow-Up

1. For the six activities you just did, combine the observations and inferences that you made in your Personal Observations and Inferences Chart with those of your classmates. Record your answers in the attached Class Observations and Inferences Chart. Place a check mark (✓) beside the best inference.

   Answer to 1: Student responses will vary; selection of the best inference will depend on class opinion.

2. Inferences often raise new questions and suggest more experiments. Consider the Curious Cup activity. What new questions might be raised by it? What experiments might you try in order to answer these questions?

   New questions could include the following: Will the experiment work if the glass is half full? How long will the paper remain on the glass? How large a glass can I use? What would happen if cardboard were used instead of paper? Students should suggest appropriate experiments to find answers to their new questions.

3. Consider this statement: Inferences are not necessarily true. Do you agree or disagree? Use the activities in this Exploration to support your answer.

   Students should agree with the statement. Most students probably made some inferences for the activities in this Exploration that they later learned were untrue.

4. What do you consider to be the value of making inferences?

   Sample answer: Making inferences helps us to make sense of what we observe. New questions and experiments are often suggested by an inference.

Name _____ Date _____ Class _____

Exploration 4 Worksheet, continued

Personal Observations and Inferences Chart

Activity	Observations	Inferences
The Dancing Disk	Student responses will vary.	Selection of the best inference will depend on class opinion. See the Answer Key for page 26.
The Reappearing Coin		
The Curious Cup		
Ice Heist		
Lights Out! #1		
Lights Out! #2		

Name _____ Date _____ Class _____

Exploration 4 Worksheet, continued

Class Observations and Inferences Chart

Activity	Observations	Inferences
The Dancing Disk	Sample answers: The disk bounces on top of the bottle.	The hands warm the bottle and the air inside it. The air expands and, as it escapes, causes the disk to move.
The Reappearing Coin	As water is poured into the dish, the coin becomes visible.	Light entering and leaving the water is bent (refracted), allowing the coin to be seen again.
The Curious Cup	When the glass is turned upside down, the water stays in the glass.	Air pressure on the outside of the paper holds it in place and keeps the water in the glass.
Ice Heist	When the string is pulled up, the ice cubes are attached to it.	The salt causes the ice to melt. The salt water refreezes, thus freezing the string to the ice.
Lights Out! #1	Bubbles may be seen escaping from the jar. When the flame is extinguished, the water level rises in the jar.	The heat from the candle expands the air, and some air escapes, causing bubbles. When the flame is extinguished, the air in the jar cools. The cooling air contracts, allowing water to enter the jar.
Lights Out! #2	When the gas in the 1 L container is "poured" over the candle, the flame is extinguished.	The gas produced in the container (carbon dioxide) is heavier (denser) than air. It flows into the jar, displaces the air, and extinguishes the candle's flame.

Name _____ Date _____ Class _____

≋ Chapter 2

Challenge Your Thinking, page 37

1. Can I Have the Recipe?

Read the two recipes below for making a blueberry pie, and then answer the following questions:

Grandma's recipe	Cookbook recipe
Simply mix three handfuls of berries with two handfuls of sugar, a handful of flour, and a pinch of salt.	Line a 23 cm pie plate with pie crust. Combine in a mixing bowl: 200 mL sugar 60 mL flour 6 mL tapioca 25 mL lemon juice Stir mixture gently into 950 mL fresh blueberries. Pour mixture into pie crust, and dot with 15 mL butter.

a. If you follow Grandma's recipe, will your pie taste the same as hers? Explain.

It will probably not taste the same because her measurements are not

precise and are therefore difficult to duplicate exactly. For example, her

"handful" and someone else's would probably be different.

b. Are Grandma's recipe and the cookbook recipe qualitative or quantitative? Explain.

Both are quantitative, but Grandma's measurements are based on one of

her handfuls, rather than a known unit of measure. To make her meas-

urements more accurate, you could measure the amount her hand can

hold in metric units. Then it is more likely that her results could be

duplicated.

Name _____ Date _____ Class _____

Chapter 2 Review Worksheet, continued

b. Suggest several inferences about what you observed.

Sample inference: The dishwashing detergent somehow causes the food coloring to move and mix with the milk.

c. Suggest one investigative question that would help you to discover more about the phenomenon.

Sample question: Does the same thing happen when water is used instead of milk?

d. Suggest a hypothesis based on your question.

Sample hypothesis: If water is used instead of milk, the same thing will happen because the food coloring and dishwashing detergent will produce the same results.

6. How Predictable!

In your own words, explain the difference between a prediction and a hypothesis. Use examples if necessary.

Sample answer: A prediction states the expected outcome of a future event, while a hypothesis provides an explanation for the outcome by linking a cause and an effect in a statement that can be tested. Sample prediction: This plant will grow more quickly than that one. Sample hypothesis: Plants given fertilizer grow more quickly than those not given fertilizer.

Name _____ Date _____ Class _____

Chapter 2 Review Worksheet, continued

2. Who Am I?

On an index card, make a table with the following headings: "Quantitative traits" and "Qualitative traits." Use this table to describe yourself. Choose one person to read the cards (but not the names) to the class so that other students can try to guess who is being described. The best description wins!

3. Thinking Like a Scientist

On page 5 of your textbook, the following statement was made: "This unit is about how you become a scientist each and every day." Do you think this is true? Support your position with examples.

Answers will vary. Most students can probably give examples of how they act like scientists in their daily lives by observing, inferring, measuring, devising experiments, and gaining new knowledge.

4. Current Events

Choose a newspaper article that has a scientific focus. Underline statements in the article that you think are inferences. Find out if a classmate agrees with you.

5. On Your Own

Try this experiment at home:

• Fill a pie plate halfway with milk.
• Add several drops of different colors of food coloring around the edge of the pie plate.
• Add several drops of dishwashing detergent to the center of the pie plate.
• Observe for 3 minutes.

a. Record as many observations as you can.

Students should observe that before the dishwashing detergent is added, the color stays in little globs. As the dishwashing detergent is added, the globs of color shoot out toward the edge of the pie plate. The colors swirl up and mix with the milk, especially where the dishwashing detergent was added.

Name _____ Date _____ Class _____

Chapter 2 Assessment, continued

c. How might you explain one of the differences you noticed in the drawings?

Since very little time has passed since the man weighed himself the

first time, it is possible that the scale is not very accurate.

d. What additional information, if any, would you like to have in order to verify your explanation?

Sample answer: I would like to verify that the scale is accurate.

Short Response

3. Read what Jorge and Julie had to say about a problem they noticed, and underline each observation and inference that they made. Then indicate in the right column whether the underlined information is an observation or an inference.

Jorge: Have you noticed how murky and brown the lake is today? — O

Julie: Yes! It looks terrible. Perhaps all the rain we've had recently has washed some extra soil into the lake. — I

Jorge: Perhaps. Or maybe more people have been visiting the lake and leaving their trash behind lately. — I

Julie: I don't think that's the problem because I noticed that the lake was this murky last month after that big rain we had. — O

Jorge: Yes, but that storm happened right after the Fourth of July weekend. This lake gets more visitors that weekend than any other time of year, and I heard that this year more people visited the lake than ever before. — O

Julie: Maybe we're both right. The large number of people using the shorelines of the lake could've trampled a lot of the plants that hold the soil in place. Then the area would've been more vulnerable to erosion. So, when the rains came, more soil was washed into the basin. — I

Jorge: Yes; and maybe the plants that hold the soil in place have still not fully recovered. Until they do recover, I bet we continue to see a murky lake after every rain. — I

Julie: I think we've got ourselves the workings of a pretty good theory here. Let's see if we can go gather some more information from the park ranger! — I

31

Name _____ Date _____ Class _____

Chapter 2 Assessment

Word Usage

1. Explain how the following pairs of terms are related:

a. *measurement* and *quantitative observations*

Quantitative observations involve measurements and numbers.

b. *hypothesis* and *cause and effect*

A hypothesis is the link between a cause and effect; a good hypothesis will indicate a cause-and-effect relationship.

Illustration for Interpretation

2. Examine the situations illustrated below for similarities and differences.

a. Compare these two situations, and write a quantitative observation about what you see.

The reading shown on the scale has increased by 1 unit in the second illustration; the time shown on the clock has increased by 10 minutes in the second illustration.

b. Compare these two situations, and write a qualitative observation about what you see.

There is a dog in the first illustration; there is a cat in the second illustration.

30 UNIT 1 • SCIENCE AND TECHNOLOGY

Chapter 3
Exploration Worksheet

EXPLORATION 1

Air and the "Paper Thing," page 43

Your goal	to investigate how certain types of objects fall through air

Safety Alert!

You Will Need

- paper
- paper clips
- scissors
- a metric ruler

What to Do

1. Make a P.T. using the attached diagram.

2. Drop your P.T., and observe its motion.

3. List all of the variables you can think of that could affect its motion.

   Some variables that could affect its motion are the height from

   which it is dropped, room temperature, size of paper clip, and type

   of paper.

4. Write one hypothesis about how you could change the way in which the P.T. falls. Exchange your hypothesis with another group.

   Sample hypothesis: A larger paper clip would make the P.T. fall

   faster.

5. Design and then try an experiment to gather evidence that either supports or contradicts the hypothesis that you received from the other group.

   Answers will vary but should be clear and complete. Experimental

   designs should test only the variable mentioned in the hypothesis.

   _____

   _____

   _____

6. Share your plan and the results with the group that gave you the hypothesis.

Chapter 2 Assessment, continued

CHALLENGE 1
Short Essay

4. Marilyn has three cats: Tiggy, Fluffball, and Boots. She wants to know which brand of cat food they like best. For one week, she feeds Brand A to Tiggy, Brand B to Fluffball, and Brand C to Boots. More of Brand C is eaten that week so Marilyn decides that Brand C must be the best.

Was Marilyn's test fair? Why or why not? If not, what could she do to improve it?

Marilyn's experiment was not a fair test. Boots may have eaten more of

Brand C than the other cats ate of their brands for a variety of reasons.

For example, Boots might be larger or more active than the other cats. To

conduct a more scientific test, Marilyn could give all three cats Brand A

the first week, Brand B the second week, and Brand C the third week.

She could then record the results and compare them to determine which

brand the cats ate the most.

CHALLENGE 2
Numerical Problem

5. Kelli has just been to a track meet. She enjoyed watching the hurdles competition and is interested in competing in the event next year, but she is not sure if she can jump high enough. Each hurdle is 91 cm tall. If Kelli is 1.73 m tall, what percentage of her height will she have to jump in order to clear each hurdle? What kind of variables would affect her success? How difficult do you think her attempt would be?

Kelli will need to jump about 53% of her height (.91 m/1.73 m =

.526 × 100 = 52.6%). Some variables that will affect Kelli's hurdling

success are her natural jumping ability, how far away from the hurdle

she is when she jumps, and how fast she approaches the hurdle. If Kelli

practices, she should be able to accomplish her goal.

Name _____ Date _____ Class _____

EXPLORATION 2

Meter Stick Experiments, page 45

Cooperative Learning Activity

Group size 4 to 5 students

Group goal to develop a hypothesis to explain reaction times

Individual responsibility Each member of your group should choose a different role such as recorder, timer, investigator, or orator.

Individual accountability Each group member should be able to prepare a data sheet that shows the group's results and include a list of any conclusions that were drawn from the results.

Experiment 1

In the first part of this experiment, you are going to compare the reaction times of the people in your group. What factors might have an effect on reaction time?

Sample factors: length of finger, age of student

What to Do

1. Person A holds the meter stick between the fingers of person B. Person A drops the meter stick, and person B catches it. The distance that the meter stick falls before person B catches it will be a measure of that person's reaction time.

2. In groups of four or five, devise rules that should be followed in order to make this a fair test. These rules should include your controlled variables. **Answers will vary but should be clear and logical. Some possible rules include the following:**

   • **Hold your fingers at the same centimeter mark each time.**

   • **Keep your fingers a certain distance from the meter stick.**

   • **Do not make any distracting noises during the trials.**

   • **Give each person the same number of trials.**

   • **Either average the results or use the best time.**

Name _____ Date _____ Class _____

Exploration 1 Worksheet, continued

Making a Better P.T.

What makes a better P.T.?

1. In making a better P.T., would you make the same one for each of the following definitions of "better"? Explain.

   • It's one that falls faster.

   • No, slower.

   • It's one that spins more often before hitting the floor.

   No; changing the definition of "better" will almost certainly change the way the P.T. is created. For example, a smaller P.T. will fall faster and spin faster. Putting a larger paper clip on the P.T. will also make it fall and spin faster. A larger P.T. may simply drop to the floor without spinning.

2. What is your definition of a better P.T.?

   Answers will vary but should be clear and logical.

3. Make your better P.T., and test it against those of your classmates. Who has the "best" P.T.?

   To make a better P.T., students might try changing the type of paper used to make the P.T. or the size or shape of the P.T., or students may create an entirely new design.

Name _____ Date _____ Class _____

Exploration 2 Worksheet, continued

Experiment 2

What to Do

1. Hold a meter stick with two fingers, as shown on page 46 of your textbook.
2. Now move your fingers toward the center of the meter stick.
3. Can you move just one finger?
4. Try this again, and carefully observe what happens. Where do your two fingers always end up?

Students should find that the two fingers always end up at the center of gravity of the meter stick. Both fingers must move toward the center of the meter stick to keep it balanced. If only one finger moves, the meter stick will eventually fall.

A Problem to Solve

1. By adding modeling clay to the meter stick, think of a way for your fingers to end up (a) at the 20 cm mark on the meter stick and (b) at the 40 cm mark on the meter stick.

Adding clay to one end of the meter stick changes the center of gravity and the finger positions.

2. Devise a hypothesis for this experiment.

Sample hypotheses: (a) If just the right amount of clay is placed at the 0 cm mark, then your fingers will end up at the 20 cm mark. (b) If one-third as much clay is placed at the 0 cm mark, then your fingers will end up at the 40 cm mark.

3. Do your results support your hypothesis? What would be an appropriate conclusion?

Students should conclude that their observations do or do not support their hypotheses.

Name _____ Date _____ Class _____

Exploration 2 Worksheet, continued

3. Test the reaction times of the members in your group using your set of rules. Record your answers in the following data chart:

Sample answers:

Trial	Distance the meter stick fell
1	18.6 cm
2	17.9 cm
3	17.5 cm
4	17.4 cm
5	16.9 cm

4. Exchange your rules with another group. Try their rules. Do you like their rules better? Answer this question by writing down how you would improve their set of rules.

Answers will vary but should be clear and should include logical recommendations.

5. Use your new, revised set of rules to test one or more of the following hypotheses:
   • Reaction time is slower when the person is seated.
   • Girls have faster reaction times than do boys.
   • Right-handed people have faster reaction times than do left-handed people.

On a separate sheet of paper, create a data chart similar to the one above to record your findings.

Name _____ Date _____ Class _____

Challenge Your Thinking, page 52

1. Words of Wisdom

Most of our everyday expressions are based on observations. Not all of these observations are accurate. Which of the following expressions do you think are true or partly true? Which could be scientifically tested?

Sample answers:

Expression	True or partly true	Testable?
• You can catch a bird by putting salt on its tail.	X	X
• You can't teach an old dog new tricks.		X
• You can catch more flies with honey than with vinegar.	X	X
• It's always darkest before the dawn.		X
• A watched pot never boils.	X	X
• There is always a calm before a storm.	X	
• A stitch in time saves nine.	X	
• A rolling stone gathers no moss.	X	
• You reap only what you sow.	X	

Illustration also on page 52 of your textbook

a. Write an investigative statement for four of the expressions above.

Some investigative questions are as follows: Does the age of a dog affect its ability to learn a new task? Are flies attracted by the smell of different substances? Is the amount of light at its lowest level just before sunrise?

_____

_____

Name _____ Date _____ Class _____

Graphing Practice Worksheet, continued

Think About It

1. The hypothesis that Kimiko and Antoine were testing was that the tighter the rubber band is wound (the more turns it is given), the longer the plane will fly.

2. Antoine and Kimiko were confused because some of their test data did not match their hypothesis, but some of it did. Looking at the graph, can you tell what data caused the problem?

   The plane flew for the same amount of time when the propeller had been given 35 and 45 turns.

3. Do you think that Kimiko's and Antoine's hypothesis was correct? Are there any other variables not shown on the graph that could have affected their hypothesis?

   Answers will vary. Generally speaking, Kimiko's and Antoine's hypothesis is correct. However, there are variables that they did not include, which could have affected their test results. Some of these include wind speed and direction, how far the plane was tossed (more important when the rubber band is not wound tightly), and whether the plane flew toward the ground at any point.

4. What sorts of variables in the design of the plane affect how long it can stay in the air? Can you think of any changes that might improve the plane's ability to stay airborne?

   Answers will vary. The two main features that students will probably mention are the size of the wings and the size of the rubber band (which is acting as the engine for the plane). Other changes might be to make the plane lighter or to try and balance the plane so that the nose stays level (preventing nose dives into the ground).

Extension

Find a partner and try Kimiko's and Antoine's experiment yourself! All you will need is a rubber-band-powered model plane, an open area, some graph paper, and a watch with a second hand. Remember to follow this simple formula: State a hypothesis, design an experiment, collect observations and data, and summarize your findings by drawing a conclusion. Good luck, and happy flying!

≋ Chapter 3

≋≋ **Answers • Chapter 3**

Name _____ Date _____ Class _____

Chapter 3 Review Worksheet, continued

4. Speak for Yourself

The title of Lesson 2 is Technology—Brainchild of Science. In your own words, describe what these words mean to you, and give examples.

Student answers will vary but could include ideas about new inventions.

Answers should also show how the knowledge gained through scientific

inquiry makes advances in technology possible.

_____

_____

_____

5. Draw Me a Map

(concept map)

Create a concept map to show how the following words are related: *future, knowledge, imagination, science, you, experiments, technology,* and *problems.* To create the map, place these words in circles, arrange them in a logical way, and link them with lines and connecting phrases.

The connections that students draw should be reasonable and thoughtful. For example, between *you* and *problems*, students may write "use science to solve."

Name _____ Date _____ Class _____

Chapter 3 Review Worksheet, continued

b. Design an experiment to test one or more of the expressions. **Answers will vary. Encourage students to use their imagination to answer this question. Remind them to control their variables. For example, several dogs of the same breed and the same age could be taught a skill. Their results could be compared to the success of younger dogs of the same breed.**

_____

2. Time Travel

Imagine being transported back in time to 1900, when the first automobiles were traveling the roads. What are some other ways in which the technology of 1900 differs from that of today?

Answers will vary, but a few examples of new inventions since 1900 are computers, televisions, radios, and sophisticated communication and transportation devices.

_____

_____

3. Life Imitates Art

Jules Verne, a famous science-fiction writer, wrote about technological advances many years before they were invented. His stories included the submarine, the helicopter, and even the fax machine. Imagine that you are a science-fiction writer and that you are writing about events in the year 2075. What new advances in technology would you include in your story?

Student answers will vary, but the obvious candidates for change are in the areas of computers, communication, and transportation. Accept all reasonable ideas.

_____

_____

≋ Chapter 3

Name _____ Date _____ Class _____

Chapter 3 Assessment, continued

4. Look at the following picture. The car on the left is typical of the cars that were built around 1910. The car on the right is a modern automobile.

Model T   Sports car

a. How are these two kinds of cars similar?

Answers will vary. Students may note that both cars have four wheels,

run on gasoline, and use steering wheels.

_____

b. How are they different? What kinds of technological advances have been made in cars since 1910?

Answers will vary. Students might mention that modern cars are more

aerodynamic, that they use a different kind of tire, and that they incor-

porate devices such as air bags, safety belts, and computers. Car

stereos and car phones are other possibilities.

c. How do you think cars might change in the future?

Answers will vary. Accept any answer that seems feasible.

_____

Name _____ Date _____ Class _____

Chapter 3
Assessment

**Correction/
Completion**

1. The statements below are incorrect or incomplete. Your challenge is to make them correct and complete.

a. The application of scientific knowledge to solve practical problems and make new inventions is _____ **technology**

b. "I think refrigeration will slow the growth of mold on bread." This is a conclusion about a cause-and-effect relationship.

• **This is a *hypothesis* about a cause-and-effect relationship.**

**Short
Response**

Airfoil A   Airfoil B

2. Martha thought that airfoil *A* would have greater lift than airfoil *B*. She is going to test her hypothesis by holding the airfoils in front of an electric fan.

a. What variable is Martha testing?

The shape of the airfoil

b. Name two variables that Martha should control.

The distance she holds the airfoils from the fan and the speed of

the fan

**Numerical
Problem**

3. Jeff wants to test three different brands of paper towels to see which absorbs the most water, which is strongest when dry, and which is strongest when wet. He will conduct each test twice to make sure he gets the same results both times. Including the repeated tests, how many tests will Jeff conduct in all?

18; 3 experiments × 3 brands × 2 trials = 18

Chapter 3

Unit 1
Unit Activity Worksheet

A Hidden Word Puzzle

Try this activity as you conclude Unit 1.

Unlock the following hidden word, and discover who is known as the "father" of modern science. His name is hidden in the clues below, which describe some of the methods of science that he used. The answers are some of the words that you have learned in Unit 1. The first one is solved for you.

① T E C H N O L O G Y
② _ _ V A R I A B L E _
③ _ M O D E L _ _
④ _ _ I N F E R R E D _
⑤ C O N T R O L L E D _
⑥ H Y P O T H E S I S _
⑦ _ _ O B S E R V A T I O N S

Clues

In 1609 he used a new example of _____ 1 _____, the telescope, to discover that the planet Jupiter has satellites, or moons, revolving around it. Previously, it was assumed that only Earth had a moon.

His discoveries of Jupiter's moons supported a new _____ 3 _____ of the solar system with the Sun instead of the Earth at its center. Although this knowledge is taken for granted now, it was a radical idea back then. Previously, having observed that the Sun appeared to move through the sky, people _____ 4 _____ that the Sun actually revolved around the Earth.

He did not rely on what others said about the nature of things but made his own careful _____ 7 _____ and conducted his own experiments. He was one of the first to do so.

He wondered if all objects, regardless of their mass, fall at the same speed. His _____ 6 _____ was that all objects fall at the same rate regardless of their mass.

One story says that he dropped iron balls from the top of the Leaning Tower of Pisa and measured the time it took for them to hit the ground. In this _____ 5 _____ experiment, the only _____ 2 _____ to change was the mass of the iron balls. His hypothesis was correct.

Chapter 3 Assessment, continued

CHALLENGE 2
Graphing Data

5. The following table compares some common methods of transportation from various times in the nineteenth and twentieth centuries:

Mode of transportation	Average speed (in kilometers per hour)
Stagecoach (1846)	13 km/h
Steam locomotive (1895)	75 km/h
Automobile (present)	100 km/h
Jet airliner (present)	1000 km/h

Using the information in the table, create a bar graph that shows how far each vehicle would travel in 1 hour. Label one axis with the names of the modes of transportation and the other axis with the distance traveled.

a. Using your graph, make one or more observations about transportation changes in the last 150 years.

Students should note that the speed of transportation has generally increased as technology has advanced. Additional responses will vary.

_____
_____
_____

b. What are some of the benefits of the advances in technology shown on the graph? Do you think there are any disadvantages associated with these changes?

Answers will vary. Some of the benefits would be faster travel, faster shipments of fresh foods and other goods, faster mail delivery, and the ability to carry many people from place to place. Disadvantages might include pollution and traffic congestion.

≈≈≈ Chapter 3

Name _____ Date _____ Class _____

Unit 1 Review Worksheet, continued

2. Below is another cryptogram. Decipher the cryptogram to discover a quotation. Hint: One of the words in the quotation is technology.

V N V W   N S Z   U H V A V U A   U K   W U Y Z
I T   I S   T H E   O P I N I O N   O F   S O M E

N S M N   O V K Z   U A   Z M F N S   J U E O B
T H A T   L I F E   O N   E A R T H   W O U L D

C Z   C Z N N Z F   V K   A U N   K U F
B E   B E T T E R   I F   N O T   F O R

W P V Z A P Z   M A B   N Z P S A U O U D L .
S C I E N C E   A N D   T E C H N O L O G Y .

After you crack the code, write a paragraph or two stating whether you agree or disagree with the quotation and why. Use a separate sheet of paper if necessary.

Answers will vary. Get students started by having them make a list of pros and cons for science and technology. For example: Technology can be used to stop or lessen pollution (pro). Technology can cause pollution (con).

3. Consider the following statement: Scientists discover; inventors invent. Explain in writing what you think this statement means. Indicate whether you agree or disagree and why.

Sample answer: Scientists and inventors share many skills, such as performing careful analysis and problem solving. While scientists usually seek new explanations for things they observe, inventors usually seek new solutions to existing problems. Student opinions will vary but should be clear and complete.

Name _____ Date _____ Class _____

Unit 1
Unit Review Worksheet

Making Connections, page 54

The Big Ideas

In your ScienceLog, write a summary of this unit, using the following questions as a guide:

1. What does it mean to be scientific? (Ch. 1)

2. What is an observation? an inference? a conclusion? (Ch. 2)

3. How does each of these help people do science? (Ch. 2)

4. How does an investigative question differ from a question that is not investigative? (Ch. 2)

5. How are the terms *hypothesis*, *variable*, and *controlled experiment* related? (Ch. 2)

6. What is the purpose of experiments? (Ch. 2)

7. What are some of the steps in designing experiments? (Ch. 2)

8. How are science and technology related? (Ch. 3)

9. How do science and technology help each other? (Ch. 3)

A sample unit summary is provided on page 54 of the Annotated Teacher's Edition.

Checking Your Understanding

1. Yvonne's group made the statements below as they did the candle activity at the beginning of this unit. Which statements are inferences? Which are observations?

a. The candle is blue. **Observation**

b. The candle is 5 cm tall. **Observation**

c. A pool of liquid forms on top of the candle as it burns. **Observation**

d. This liquid is made of the same substance as the candle. **Inference**

e. The candle flickers as it burns. **Observation**

f. Blowing hard on the candle causes it to go out. **Observation**

g. Blowing hard on a candle causes it to go out because you blow all of the air away from it. **Inference**

h. Candles need air to burn. **Inference**

Name _____ Date _____ Class _____

Word Usage

1. The following words belong together. For each pair, discuss why they belong together.

   a. *science* and *discoveries*

      Sample answer: The purpose of *science* is to make *discoveries* about the world around us.

   b. *observation* and *inference*

      Sample answer: An *inference* may explain an *observation*.

Correction/Completion

2. The statements below are incorrect. Your challenge is to make them correct.

   a. In talking about observations, Robert said, "Observations are what you see."

      In talking about observations, Robert said, "Observations *can be made with any of the senses*."

   b. In a controlled experiment, every variable is controlled.

      In a controlled experiment, every variable *except one* is controlled.

   c. "Marie has brown eyes and freckles." These are quantitative observations.

      "Marie has brown eyes and freckles." These are *qualitative* observations.

Short Responses

3. Classify these statements as either observations or inferences.

   a. The red appearance of a sunset is caused by the blanket of air through which the sun is shining.   **Inference**

   b. The noonday sun is warmer in the summer than it is in the winter.   **Observation**

   c. The sun appears larger as it is setting than it does at noon.   **Observation**

   d. The Sun and the Earth are the same age.   **Inference**

Name _____ Date _____ Class _____

Unit 1 Review Worksheet, continued

4. Once upon a time, there was a young boy who lived in the country. This boy noticed that every morning, just before dawn, the roosters began to crow. He hypothesized that the roosters' crowing caused the sun to rise. Design an experiment to test this hypothesis.

   Answers will vary, but students should indicate that the variable to change is the time at which the rooster crows.

5. If an experiment repeatedly disproves a hypothesis, which of the following actions would be a correct response for a scientist?

   a. Ignore the results of the experiment.

   b. Keep trying new experiments until the hypothesis is supported.

   c. Reject the old hypothesis and form a new one.

   d. Conclude that the experiment had some sort of flaw.

   Justify your response in writing.

   A scientist should make sure that his or her experiment is a good one. New experiments to test the hypothesis can be tried, but eventually new hypotheses may need to be formed. Therefore, choices (c) or (d) would be correct.

6. Complete the concept map below by using the following words: *the world, observing, scientists, inferring, experimenting, variable, hypotheses,* and *models*.

   Sample concept map:

Photo also on page 55 of your textbook

Name _____ Date _____ Class _____

Unit 1 Assessment, continued

4. Identify the cause and effect in these hypotheses.

   a. The higher the temperature of the water, the faster the eggs will cook.

   Cause: higher water temperature; effect: faster rate of cooking

   b. People who drink fluoridated water will have less tooth decay than those who don't.

   Cause: drinking fluoridated water; effect: less tooth decay

5. You are observing water flowing from a kitchen tap.

   a. Suggest three observations that you could make.

   Sample observations: The water is flowing rapidly; the water is extremely hot; the water has been running for at least several minutes.

   b. Write down one inference.

   Sample inference: Someone is about to wash the dishes.

Data for Interpretation

6. Chris and Tammy recorded the following results after flipping a coin 10 times:

Test	Results
1	tails
2	tails
3	heads
4	heads
5	tails
6	tails
7	heads
8	tails
9	heads
10	tails

   a. What percentage of the time did the coin turn up heads? **40%**

Name _____ Date _____ Class _____

Unit 1 Assessment, continued

   b. If they flipped the coin 100 times, what percentage of the time do you predict the coin would turn up heads?

   Sample answer: 50%

   c. What investigative question could Chris and Tammy have been trying to answer?

   Sample answer: What percentage of the time will a coin turn up heads?

CHALLENGE

Short Response

7. Another way of describing technology is to use the term *applied science.* Explain why this is a good synonym for technology.

   Sample answer: Technology puts the theories and concepts of science into action. Technology applies the discoveries of researchers and scientists to inventions, industry, and everyday uses. Thus, technology can also be called *applied science.*

Data for Interpretation

8. Justine recorded the following data in her ScienceLog.

Amount of sugar in water	Time to boil
0 g	5 min. 53 sec.
6 g	6 min. 5 sec.
12 g	6 min. 17 sec.
18 g	6 min. 29 sec.
24 g	?

   a. What investigative question is Justine trying to answer?

   Sample answer: How does adding sugar to water affect the amount of time it takes for the water to boil?

Name _____ Date _____ Class _____

Unit 1 Assessment, continued

Graph for Interpretation

10. The graph below represents data from an experiment. The graph shows how long a music box played after different numbers of turns of the key.

a. What hypothesis do you think is being tested?

The more times the key of a music box is turned, the longer it will play.

b. What should have been done to ensure that the experiment was a fair one?

The same music box should be used for each trial. Each trial should begin with the same amount of tension on the spring.

c. What would be a good conclusion for this experiment?

The amount of time the music box plays is not directly proportional to the number of turns of the key. Past a certain point, the number of turns does not make much difference in the length of play.

Name _____ Date _____ Class _____

Unit 1 Assessment, continued

b. What variables need to be controlled to make this a fair test?

Amount of water, starting temperature of water, amount of heat applied

c. How long do you predict that it would take to boil water with 24 g of sugar in it?

6 min. 41 sec.

CHALLENGE 2
Short Essay

9. Jill heard an ad on television claiming that a certain shampoo cleaned hair better than any other brand. She decided to do an experiment to test this claim. Describe the experiment she might do. What variables must be controlled to make it a fair test?

Answers will vary but variables that would have to be controlled include the following: the dirtiness of the hair, the hair length, and the type of hair (thin, thick, dry, oily, etc.). An objective method for measuring the cleanliness of the hair would also have to be devised.

Unit 1

Name _____ Date _____ Class _____

Unit 1
Activity Assessment

Thinking Scientifically

As a scientist, your job is to answer investigative questions. You get to make observations, form hypotheses, and experiment to test your hypotheses. Show how scientific thinking and sharp science skills can help you accomplish the tasks below.

Before You Begin . . .

As you work through the tasks, keep in mind that your teacher will be observing the following:

• how you use the materials and equipment

• how clear and complete your observations are

• how well you draw conclusions from your observations

Now you are ready to tackle the tasks!

Task 1: Mysterious Contents

Without opening the closed container, find out as much as you can about the object inside. You may use any of the materials provided. Record your observations and inferences below.

1. Qualitative observations:

   Answers will vary depending on the object used. Observations should be

   clear and concise and should not involve measurements or numbers.

2. Quantitative observations:

   Answers will vary depending on the object used. Observations should be

   clear and concise and should involve measurements or numbers.

3. Inferences:

   Answers should show the student's ability to draw a logical conclusion

   from his or her observations.

4. In what ways have you been scientific in your study of the unknown object?

   Sample answer: By making observations and inferences to learn some-

   thing about an unknown object, I have been scientific in my study.

Name _____ Date _____ Class _____

Activity Assessment, continued

Task 2: Sand and Water

Make as many observations as you can about the containers of sand and water.

1. Qualitative observations of container of water:

   Sample answer: The container is made of a sturdy, red plastic. The water

   is clean and cool to the touch. It flows when I move the container.

2. Qualitative observations of container of sand:

   Sample answer: The container is made of a sturdy, blue plastic. The sand

   is mostly white, and it flows when I move the container.

3. Quantitative observations of container of water:

   Sample answer: The temperature of the water is 27°C. The container with

   the water weighs 135 g.

4. Quantitative observations of container of sand:

   Sample answer: The temperature of the sand is 23°C. The container with

   the sand weighs 165 g.

Task 3: Water + Sand = ?

Now pour the water onto the sand and make further observations.

1. Qualitative observations of container of water and sand:

   Sample answer: The sand is now thoroughly soaked, and a thin layer of

   water covers its surface. The sand has turned a much darker color and is

   well packed instead of loose.

2. Quantitative observations of container of water and sand:

   Sample answer: The temperature of the sand-and-water mixture is 26°C.

   The container with the mixture weighs 275 g.

3. Inferences:

   Sample answer: When sand gets wet, it no longer flows but can be easily

   molded into various shapes. Wet sand is also cooler and heavier than dry

   sand.

4. In what ways have you been scientific in this study?

   Sample answer: I made observations and inferences about two substances and then

   combined them to learn something about a third substance.

Name _____ Date _____ Class _____

SourceBook Activity Worksheet, continued

Experimental Design Form

Hypothesis
The carnations turned pink because they absorbed something that was in the water.

Plan
Experimental plans should be clear and complete. Be sure that students identify a control and consider several variables. Sample plan: I could put one white carnation in a beaker with 300 mL of water. This would be my control. Then I could set up three other beakers, *A*, *B*, and *C*, and put 300 mL of water in each. To beaker *A*, I could add a small quantity of food coloring; to beaker *B*, I could add a small quantity of chili powder; and to beaker *C*, I could add some water-soluble red paint. Then I could watch what happened. If nothing happened immediately, I could continue making observations over the next 3 days.

Results (Observations)
Observations should be clear and concise. If students do not achieve successful results at first, encourage them to start the process over again with a new hypothesis. Sample observations: The control carnation stayed white. The carnation in beaker *A* turned pink, the carnation in beaker *B* wilted slightly, and the carnation in beaker *C* wilted dramatically.

Conclusion
Sample conclusion: Since the flowers in the classroom did not wilt but did turn pink, we conclude that Craig must have added food coloring to the water.

Name _____ Date _____ Class _____

≋ Sourcebook

Unit CheckUp, page S21

Concept Mapping

The concept map shown here illustrates major ideas in this unit. Complete the map by supplying the missing terms. Then extend your map by answering the additional question below.

Where and how would you connect the terms *entomology, astrophysics,* and *drawing conclusions?*

Name _____ Date _____ Class _____

SourceBook Review Worksheet, continued

Checking Your Understanding

Select the choice that most completely and correctly answers each of the following questions.

1. An Earth scientist would most likely investigate which of the following?
   a. (an underwater volcano)
   b. an African tree lizard
   c. an unbalanced nuclear force
   d. an insect invasion

2. A scientific method is a
   a. technique for finding answers.
   b. process for doing experiments.
   c. (way of thinking about the natural world.)
   d. reliable system for recording observations.

3. A *control* for any experiment is necessary because
   a. the scientific method requires one.
   b. it predicts what the results should be.
   c. (it establishes a basis for comparison.)
   d. scientists want to be in charge.

4. If an experiment supports a hypothesis,
   a. the hypothesis is valid.
   b. (the hypothesis may be valid.)
   c. the hypothesis is invalid.
   d. the hypothesis is incomplete.

5. When scientists fail to communicate their conclusions,
   a. they get all of the glory for themselves.
   b. their work cannot be considered scientific.
   c. (other scientists cannot verify their results.)
   d. other scientists must steal their ideas.

Name _____ Date _____ Class _____

SourceBook Review Worksheet, continued

≋≋ **SourceBook**

Interpreting Graphs

Look at the graph below. What general relationship between temperature and rainfall could you infer from the information given?

As temperature rises, rainfall decreases. As temperature falls, rainfall

increases.

_____

At what point in a scientific method would someone produce a graph such as this? Explain.

This graph might be produced after a scientist made observations of the

weather and collected the data. From this graph, the scientist could then

draw conclusions about the relationship between temperature and rainfall.

Rainfall (cm)   Temperature (°C)

J F M A M J J A S O N D

Graph also on page S22 of your textbook.

Name _____ Date _____ Class _____

SourceBook Review Worksheet, continued

4. Explain why scientists must have at least two samples when they run any experiment.

There must always be a control as well as a variable. One sample must be kept unchanged while the other is adjusted in order to observe the effect of the change.

_____

_____

5. A scientific law might state that the sun will rise tomorrow morning. How might this law be affected if, for some cosmic reason, the Earth stopped rotating?

The law would have to be changed. A new description of how the Earth rotates would then need to be determined.

_____

_____

_____

_____

Portfolio Idea

Imagine that the principal of your school called you to the main office and asked you to explain the meaning of science. What would you say? Think about the science classes you have taken. Would you call what happened in them "science" or not? Make an outline to help you remember the main points you want to bring up. Remember that what you tell the principal may influence how science is taught in your school. Use another piece of paper if necessary.

Answers will vary. However, responses should include an understanding that science is something people do and that students can be a part of it.

_____

_____

_____

SCIENCEPLUS • LEVEL GREEN 67

Name _____ Date _____ Class _____

Critical Thinking

SourceBook Review Worksheet, continued

Carefully consider the following questions, and write a response that indicates your understanding of science.

1. Would the question, How can we use hydrogen to run our automobiles? be a question for science or for technology? Explain.

Technology. The use of scientific knowledge for the development of products useful to human society is called technology.

_____

_____

2. Some people argue that astronomy is *not* an Earth science. Why might they say this?

Since astronomy deals with objects other than the Earth itself, it seems strange to call it an Earth science.

_____

_____

_____

3. Why is it necessary for scientists in different fields, such as biology and physics, to communicate and work together?

Many phenomena cannot be studied in relation to a single scientific discipline. When studying the human body, for example, biologists will focus on the organ systems, while physicists study the motion of the bones.

_____

_____

66 UNIT 1 • SCIENCE AND TECHNOLOGY

104 UNIT 1 • SCIENCE AND TECHNOLOGY

Name _____ Date _____ Class _____

Unit 1 SourceBook Assessment

1. Which of the following *best* describes what science is?
   a. Science is the observation of our natural world.
   b. Science is performing experiments to find solutions to problems.
   c. (Science is the asking and answering of questions in order to satisfy curiosity about the natural world.)
   d. Science is the designing and carrying out of experiments.

2. Which of the following is the *best* description of a scientist?
   a. a person who wears a white coat and works in a laboratory
   b. a doctor
   c. a person who studies endangered animal species in the wild
   d. (a person who asks questions and is curious about the natural world)

3. If you were a biologist, which of the following would you be *least* likely to study?
   a. life cycles in the tropical rain forest
   b. how plants make food
   c. AIDS research
   d. (how earthquakes and volcanic eruptions are predicted)

4. A scientific method is only used by scientists.
   a. true   b. (false)

5. A scientific method is a technique for always getting the right answer to a problem.
   a. true   b. (false)

6. Which of the following is *not* true of a hypothesis?
   a. Each hypothesis must be tested.
   b. (Only one hypothesis can be formed at a time.)
   c. A hypothesis should be stated before an experiment is done.
   d. Scientists use observations and experiments to test a hypothesis.

7. The steps of a scientific method are not the same for all scientists.
   a. (true)   b. false

8. Which of the following has the greatest chance of being incorrect?
   a. a scientific theory   b. a scientific law   c. (a hypothesis)

Name _____ Date _____ Class _____

≋ SourceBook

SourceBook Assessment, continued

9. Scientists observed the following curious events. For each event, give a question that you think scientists might ask about that event.
   a. A large community of organisms was observed living around a crack in the ocean floor. This crack was the vent for a hot spring.
   Sample question: **Where do the organisms living around the crack in the ocean floor get their food?**

   b. The eruption of Mount St. Helens on May 18, 1980, turned the surrounding area into a landscape resembling that of the moon.
   Sample answer: **Will plants and other living things such as those that once lived here ever return?**

   c. Some rain that falls has an abnormally high acidity. Scientists call this acid rain.
   Sample answer: **What is causing this rain to be so acidic?**

10. What do we call it when scientists apply the knowledge they have gained to improve the quality of life?
    a. experimentation   b. (technology)   c. meteorology   d. science

11. Choose one of the following specific areas of science: botany, genetics, geology, meteorology, or biochemistry. Imagine that you are a scientist in this field. If someone asked you about your work, how would you describe it to him or her?
    Sample answer: **Botany is the study of plants and the conditions in which they live.**

12. Why do scientists study a specific area of science, like geology, biochemistry, or zoology, instead of studying all areas of science?
    By concentrating on one specific subject, scientists are able to learn a great deal more about that subject. Then scientists can share information with each other to learn more about the big picture.

≋ Answers • SourceBook

Name _____ Date _____ Class _____

SourceBook Assessment, continued

13. What are the two main reasons for doing scientific research?

Two main reasons are to find solutions to problems and to find out more about something.

14. Explain why a hypothesis is sometimes called an educated guess.

A hypothesis is sometimes called an educated guess because it involves some prior knowledge of the problem.

15. If an experiment is designed to test a hypothesis and it proves that the hypothesis is incorrect, was the experiment a waste of time? Explain your answer.

No. An incorrect hypothesis is valuable because it eliminates an incorrect answer and encourages new questions.

16. Explain why scientists can change only one variable at a time in an experiment.

If more than one variable at a time is changed, scientists have no way of knowing which one affected the results of the experiment.

17. Based on the results of their study, what did Bekoff and Wells discover about why coyotes run in large packs?

They learned that coyotes run in large packs when the winter supply of large, dead animals is plentiful.

18. Distinguish between a scientific theory and a scientific law.

A scientific law describes only what happens during a certain event. A scientific theory, on the other hand, describes why something happens.

Name _____ Date _____ Class _____

SourceBook Assessment, continued

19. What is meant by the statement, Science moves ahead by correcting errors it made earlier?

Answers will vary. Answers should include the understanding that scientific knowledge is always changing.

20. Form a hypothesis for a question you are curious about, and use the scientific method you learned about in this SourceBook unit to show how you could test that hypothesis.

Answers will vary, but students should first define a problem, form a hypothesis, note the need to make observations, and outline a plan for experimentation. Then they should note the need to analyze data, draw conclusions, and communicate the results.

21. Match the major areas of science on the left with the more specific areas of science on the right. (Each major area will be used twice.)

a. life science	b	geology
b. Earth science	c	sonar
c. physical science	b	meteorology
	a	genetics
	a	ecology
	c	lasers

22. Match the specialized field of science on the left with the correct object of study on the right.

a. botany	e	oceans of the world
b. geology	d	makeup of matter and its interactions
c. biochemistry	g	animals
d. chemistry	b	rocky surfaces and the interior of the Earth
e. oceanography	a	plants
f. meteorology	f	Earth's atmosphere
g. zoology	h	energy, its changes, and its relationship to matter
h. physics	c	chemistry of life

Name _____ Date _____ Class _____

SourceBook Assessment, continued

23. Match the step of a scientific method on the left with the examples on the right.

a. defining the problem

b. forming hypotheses

c. making observations

d. analyzing data

___c___ watching the social behavior of ants

___d___ graphing information collected in a recent experiment

___a___ a large number of dolphins dying this year

___b___ high nitrogen levels in the lake causing fish to die

24. The study of the universe beyond the Earth is known as

astronomy .

25. Any information we gather by using our senses is a(n)

observation .